AI for Climate Change and Environmental Sustainability

This book discusses the adverse effects of climatic changes on our planet. It examines AI-based tools and technologies and how they can assist in identifying energy emission reductions, CO_2 removal, and support the development of greener transportation networks, monitoring deforestation, and forecasting extreme weather events.

AI for Climate Change and Environmental Sustainability identifies and discusses in detail the importance of environmental sustainability based on accomplishment of the UN's 17 Sustainable Developmental Goals (SDGs). It presents the various AI-based possibilities for accelerating international efforts to safeguard the environment and conserve natural resources. The authors offer a comprehensive analysis of the emerging field of climate change in relation to Internet of Things, artificial intelligence, machine learning, and deep learning. The book discusses AI developments, applications, and best practices that will help us transition to a low-carbon future on both a regional and global scale. It provides case studies with analytical results pertinent to climate change and weather prediction and includes chapters with a research-oriented approach, which can encourage new developments in the field of sustainable climate and green environment.

The book can be used as a primary textbook for graduate and postgraduate students in technology and science, as well as a reference for researchers, academics, and IT professionals working on climate change and sustainability initiatives.

Artificial Intelligence for Sustainable Engineering and Management

Sachi Nandan Mohanty
School of Computer Science & Engineering (SCOPE), VIT-AP University, Amaravati, Andhra Pradesh, India
Deepak Gupta

Artificial intelligence is shaping the future of humanity across nearly every industry. It is already the main driver of emerging technologies like big data, robotics and IoT, and it will continue to act as a technological innovator for the foreseeable future. Artificial intelligence is the simulation of human intelligence processes by machines, especially computer systems. Specific applications of AI include expert systems, natural language processing, speech recognition and machine vision. The future of business intelligence combined with AI will see the analysis of huge quantities of contextual data in real time. So, the tool will quickly capture customer needs and priorities and do what is needed.

AI for Climate Change and Environmental Sustainability
Edited by Suneeta Satpathy, Satyasundara Mahapatra, Nidhi Agarwal, and Sachi Nandan Mohanty

Green Metaverse for Greener Economies
Edited by Sukanta Kumar Baral, Richa Goel, Tilottama Singh, and Rakesh Kumar

Healthcare Analytics and Advanced Computational Intelligence
Edited by Sushruta Mishra, Meshal Alharbi, Hrudaya Kumar Tripathy, Biswajit Sahoo, and Ahmed Alkhayyat

www.routledge.com/AI-for-Sustainable-Engineering-and-Management-series/book-series/AISEM

AI for Climate Change and Environmental Sustainability

Edited by
Suneeta Satpathy, Satyasundara Mahapatra,
Nidhi Agarwal, and Sachi Nandan Mohanty

CRC Press
Taylor & Francis Group
Boca Raton London New York

CRC Press is an imprint of the
Taylor & Francis Group, an **informa** business

Designed cover image: © Shutterstock

First edition published 2025
by CRC Press
2385 NW Executive Center Drive, Suite 320, Boca Raton FL 33431

and by CRC Press
4 Park Square, Milton Park, Abingdon, Oxon, OX14 4RN

CRC Press is an imprint of Taylor & Francis Group, LLC

© 2025 selection and editorial matter, Suneeta Satpathy, Satyasundara Mahapatra, Nidhi Agarwal, and Sachi Nandan Mohanty; individual chapters, the contributors

Library of Congress Cataloging-in-Publication Data
Names: Satpathy, Suneeta, editor.
Title: AI for climate change and environmental sustainability / edited by
   Suneeta Satpathy [and 3 others].
Description: First edition. | Boca Raton, FL : CRC Press, 2024. | Series: AI for sustainable
   engineering and management | Includes bibliographical references.
Identifiers: LCCN 2024002940 | ISBN 9781032589732 (paperback) |
   ISBN 9781003452393 (ebook)
Subjects: LCSH: Climatic changes—Technological innovations. | Artificial intelligence.
Classification: LCC QC903 .A146 2024 | DDC 363.700285/63—dc23/eng/20240521
LC record available at https://lccn.loc.gov/2024002940

ISBN: 978-1-032-58906-0 (hbk)
ISBN: 978-1-032-58973-2 (pbk)
ISBN: 978-1-003-45239-3 (ebk)

DOI: 10.1201/9781003452393

Typeset in Times
by Apex CoVantage, LLC

Contents

Chapter 8   Optimal Dispatch of Distributed Renewable Energy Sources in Isolated Microgrid System Exploiting Metaheuristic Optimization Algorithms... 98

Anirudh Kumar Verma, Prakash M and Shakila B

Chapter 9   Leveraging Machine Learning Models for Intelligent Hazard Management.. 109

Preethi Nanjundan, George Jossy P, and Karpagam C

About the Editors

Dr Suneeta Satpathy (Senior member, IEEE) is currently working as an associate professor, Center for AI & ML, SoA University, Bhubaneswar, Odisha. She received her Ph.D. from Utkal University, Bhubaneswar, Odisha, in 2015, with a Directorate of Forensic Sciences, MHA scholarship from the Government of India. Her research interests include Computer Forensics, Cyber Security, Data Fusion, Data Mining, Big Data Analysis, and Decision Mining. She has edited books in association with Springer and Wiley and CRC AAP, NOVA Publications. In addition to research, she has guided many post-graduate and graduate students. She has published papers in many international journals and reputable conferences. Her professional activities include roles as an editorial board member and/or reviewer of *Journal of Engineering Science, Advancement of Computer Technology and Applications, Robotics and Autonomous Systems* (Elsevier), *Computational and Structural Biotechnology Journal* (Elsevier), *Journal of Big Data* (Springer) as well as Inder Science Journals. She is an active member of CSI, ISTE, OITS, IE, Nikhil Bharat Shiksha Parisad and IEEE.

Dr Satyasundara Mahapatra is working as a professor in the department of Computer Science and Engineering, Pranveer Singh Institute of Technology, Kanpur, Uttar Pradesh. He received his Ph.D. and master's degree in computer science from Utkal University, Bhubaneswar, Odisha in 2006 and 2016, respectively. He also holds an MCA degree from Utkal University, Bhubaneswar, Odisha in 1997. His current research interests include Machine Learning, Deep Learning, Image Processing, Scheduling with Optimization, and Internet of Things. He has a number of publications as authored or co-authored in various international scientific journals, book chapters and five patent approvals and two patent grants in his field of expertise. He was also selected for the award of grant under "Visvesvaraya Research Promotion Scheme" for academic year 2019–2020 for his research project.

Ms Nidhi Agarwal is a computer science and engineering faculty member with 19+ years of experience, an author of many papers in SCI/SCI-E indexed journals, SCOPUS-indexed journals and conferences, books and book chapters, patents, strong subject knowledge, reviewer and editorial board member of many international journals including SCI/E, SCOPUS, session chair in many SCOPUS-indexed international conferences, and NTA NET qualified in Computer Science and Applications. She was a National Merit Scholarship Holder for outstanding performance in mathematics in class 10. She received a Research Excellence award of Rs 1 lakh in 2023 for excellence in research in 2022 from IGDTUW, Delhi. She received the CV Raman Award at KIET Group of Institutions in 2022 for being the best researcher at the institute level for the academic session 2021–22. She also received the "Shikshak Samman Award 2022" by Nikhil Bharat Shiksha Parishad, Govt. of India, for outstanding work in research.

Dr Sachi Nandan Mohanty was recognized as Top 2% World Scientists Ranking by Stanford University and Elsevier for the year 2023. He received his PostDoc from Indian Institute of Technology Kanpur, India, in the year 2019 and Ph.D. from Indian Institute of Technology Kharagpur, India, in the year 2015, with MHRD scholarship from Govt of India. He has authored/edited 42 books, published by IEEE-Wiley, Springer, Wiley, CRC Press, NOVA, and De Gruyter. His research areas include Data mining, Big Data Analysis, Cognitive Science, Fuzzy Decision Making, Brain-Computer Interface, Cognition, and Computational Intelligence. Prof. S N Mohanty has received four Best Paper Awards during his Ph.D. from International Conference at Beijing, China, and the other at International Conference on Soft Computing Applications organized by IIT Roorkee in the year 2013. He was awarded the Best thesis award first prize by Computer Society of India in the year 2015. He has guided nine PhD Scholars, 23 postgraduate students. He has published 241 International Journals of International repute. He has received fourth time International Travel support from Government of India's Department of Science and Technology, SERB Funding for International travel for presenting research paper, and Keynote specker. He has been awarded by many awards and fellowship during his career and to name a few are Prof. Ganesh Mishra Memorial Award on 61st Annual Technical Session & 19th Prof. Bhubaneswar Behera Lecturer on 23–24 June 2020 from The Institution of Engineers (India) Odisha State Center, Bhubaneswar. He is a member of professional bodies like IETE, CSI, IE, IEEE (Hyderabad section). He is also the reviewer of *Journal of Robotics and Autonomous Systems* (Elsevier), *Computational and Structural Biotechnology Journal* (Elsevier), *Artificial Intelligence Review* (Springer), *Spatial Information Research* (Springer). He is leading as General chair of two international conferences such as ICISML, AIHC and editor in chief of *International Journal EAI Transaction on Intelligent System and Machine learning Application*. Dr. Mohanty has gone for academic assignment to New York, Atlanta, Orlando, Georgia, Paris, Slovakia, Singapore, Abu Dhabi, Sharjah, Dubai, Malaysia, Istanbul, Vienna, and Germany for delivering Keynote talk and chair the sessions in international conferences with travel support from Department of Science and Technology, Government of India, New Delhi, India.

Contributors

Shakila B
Department of Electrical and
    Electronics Engineering, NIT
    Nagaland, Chumoukedima, India

Asiya Begum
School of Computer Science and
    Engineering, VIT-AP University,
    Amaravati, India

Sarthak Dash
Berhampur University, India

Deepak Gupta
Pranveer Singh Institute of Technology,
    Kanpur, Uttar Pradesh, India

Senthil Janarthanan
Bharathidasan University,
    Tiruchirappalli, Tamil Nadu, India

Naresh Kumar Kar
GITAM Deemed to be University,
    Hyderabad, India

Shaik Kareemulla
School of Computer Science and
    Engineering, VIT-AP University,
    Amaravati, India

Syed Khasim
School of Computer Science and
    Engineering, VIT-AP University,
    Amaravati, India

Pradumn Kumar
Department of Computer Science
    and Engineering, Pranveer Singh
    Institute of Technology.
    Kanpur, Uttar Pradesh, India

Rakesh Kumar
Department of Electrical and
    Electronics Engineering, NIT
    Nagaland, Chumoukedima, India

Jai Prakash Kushwaha
IIITM, Gwalior, India

Prakash M
Department of Electrical
    and Electronics Engineering,
    NIT Nagaland, Chumoukedima,
    India

Nabanita Mandal
Thadomal Shahani
    Engineering College,
    Mumbai, India

Preethi Nanjundan
Department of Data Science, Christ
    University, India

George Jossy P
Department of Data Analytics,
    Dr N.G.P. Arts and Science College,
    India

Manoj Kumar Pandey
Pranveer Singh Institute of
    Technology, Kanpur, Uttar Pradesh,
    India

Sugyanta Priyadarshini
KIIT Deemed to be University,
    Bhubaneshwar, India

Sukanya Priyadarshini
KIIT Deemed to be University,
    Bhubaneshwar, India

Priyadharshini Ravi
Bharathidasan University,
    Tiruchirappalli, Tamil Nadu,
    India

Shaik Salma
School of Computer Science and
    Engineering, VIT-AP University,
    Amaravati, India

Sunil Kumar Singh
CMREC, Hyderabad, India

Hussain Syed
School of Computer Science and
    Engineering, VIT-AP University,
    Amaravati, India

Sarode Tanuja
Thadomal Shahani Engineering
    College, Mumbai, India

Lijo Thomas
Department of Data Analytics, Dr N.G.P.
    Arts and Science College, India

Preeti Tiwari
GITAM Deemed to be University,
    Hyderabad, India

Jyoti Upadhyay
GD Rungta College of
    Science and Technology,
    Bhilai, India

Anirudh Kumar Verma
Department of Electrical and
    Electronics Engineering,
    National Institute of Technology,
    Nagaland, Chumoukedima,
    Nagaland, India

Preface

In an era defined by the urgent need for environmental solutions, this book, titled *AI for Climate Change and Environmental Sustainability*, serves as a crucial exploration into the amalgamation of artificial intelligence (AI), Internet of Things (IoT), and machine learning. As we stand at the crossroads of ecological danger, these innovative technologies emerge as powerful tools to not only understand the intricate challenges posed by climate change but also to formulate proactive strategies for environmental conservation. This book sets the stage for a profound journey through twelve chapters, each delving into distinct aspects of how AI and related technologies can pave the way for a more sustainable and resilient future.

1 AI as Sustainable and Eco-Friendly Environment for Climate Change

Preethi Nanjundan and Lijo Thomas

1.1 INTRODUCTION

Innovative and practical solutions are needed to address the urgent problems of climate change and environmental sustainability in our quickly changing world. The need to address these issues has never been more pressing as the effects of climate change, including extreme weather events, habitat loss, and resource shortages, continue to worsen. In light of this, artificial intelligence (AI) stands out as a crucial instrument that has the power to completely transform our efforts to combat climate change and promote a sustainable, environmentally friendly society.

Without a doubt, the Intergovernmental Panel on Climate Change (IPCC) has emphasised how urgent it is to stop global warming in order to avoid disastrous effects on the environment [1]. With its ability to provide data-driven insights and disruptive technologies, AI is well-positioned to be a key player in resource optimisation, conservation, and efforts to mitigate climate change. AI-driven solutions are at the vanguard of scientific and technical breakthroughs, offering fresh methods to solve these important concerns as we negotiate the intricate and interconnected difficulties of climate change.

The goal of this study is to present a thorough and in-depth investigation of AI's multidimensional contribution to sustainability advancement and climate issues. We explore the uses of AI in resource management, the adoption of renewable energy sources, environmental preservation, and climate prediction, drawing on a multitude of research and studies [2]. This study highlights the revolutionary potential of AI in altering our approach to environmental sustainability by synthesising important concepts.

In order to prevent catastrophic effects for our world, the Intergovernmental Panel on Climate Change (IPCC) has unambiguously emphasised the need of curbing global warming [3]. With its capacity to analyse enormous datasets, spot trends, and make predictions, AI is poised to fundamentally alter how we approach the fight against climate change. AI-driven solutions have become an essential part of the larger plan to handle the complex problems caused by climate change as we struggle with them [4].

DOI: 10.1201/9781003452393-1

1.2 BASIC CONCEPTS

Two key concerns that have grown more important in recent years are dealing with climate change and ensuring environmental sustainability. Long-term changes in Earth's temperature, precipitation, and other weather patterns are referred to as climate change. Human actions, especially the release of greenhouse gases like carbon dioxide (CO_2), which trap heat in the Earth's atmosphere and cause global warming, are the main cause. Significant and varied effects of climate change include increased frequency and severity of weather events, increasing sea levels, habitat loss, and changes to ecosystems.

The revolutionary potential of artificial intelligence (AI) [5] is harnessed to tackle these issues and advance the vision of a sustainable and eco-friendly world. AI, encompassing a broad domain within computer science, aims to create machines with the capacity to learn, reason, solve problems, and engage in decision-making— functions historically within the human domain. AI systems, equipped with powerful algorithms and processing capabilities, analyse vast datasets, identify trends, and generate predictions.

AI transforms into a potent instrument to optimise and improve numerous facets of environmental management when used to combat climate change and promote sustainability. This covers eco-friendly technology development, resource conservation, environmental monitoring, and energy efficiency.

1.3 AI FOR SUSTAINABLE AND ECO-FRIENDLY ENVIRONMENT

AI may play a variety of roles in combating climate change and promoting a sustainable environment. It has the ability to revolutionise how we approach these problems by presenting fresh ideas and data-driven understandings.

Energy efficiency is one important area where AI excels. AI-driven solutions can reduce energy use in commercial buildings, industrial settings, and transportation. By reducing waste and greenhouse gas emissions, this optimisation helps to create a more environmentally friendly and sustainable energy ecology. Predictive maintenance reduces energy losses in industrial equipment, whereas smart grids, for example, use AI to manage energy supply and demand effectively.

In addition, AI is crucial in agriculture, one of the most important industries for sustainability. AI-driven precision agriculture improves farming practises by using sensors and data analytics [6]. It improves crop yields while reducing resource usage and environmental effect by optimising crop management, resource allocation, and pest control.

Artificial intelligence also dramatically improves conservation efforts. Large datasets may be analysed by machine learning algorithms to monitor and safeguard animals and natural environments, such as those found in satellite photos and auditory recordings. This device helps protect endangered animals, stop illicit poaching, and preserve biodiversity.

AI encourages the use of renewable energy sources like solar and wind power. Renewable energy sources become more viable and sustainable as a result of optimising energy production, storage, and delivery.

Additionally, AI is essential for managing disasters and predicting the weather. In order to increase disaster preparedness and resilience, machine learning algorithms may examine past weather data to provide more accurate climate forecasts. This enables early warnings for extreme weather occurrences like hurricanes and floods [7].

In our attempts to build a more environmentally friendly and sustainable world, AI is a potent instrument that offers creative ways to address climate changes and promote sustainability. Its capacity for data analysis, forecasting, and resource optimisation places it in a prime position to advance in these crucial fields.

1.3.1 AI-Enhanced Climate Monitoring Systems

Additionally, AI is essential for managing disasters and predicting the weather. In order to increase disaster preparedness and resilience, machine learning algorithms may examine past weather data to provide more accurate climate forecasts. This enables early warnings for extreme weather occurrences like hurricanes and floods [7].

In our attempts to build a more environmentally friendly and sustainable world, AI is a potent instrument that offers creative ways to address climate changes and promote sustainability. Its capacity for data analysis, forecasting, and resource optimisation places it in a prime position to advance in these crucial fields.

1. Data Analysis and Pattern Recognition: Artificial intelligence (AI) makes it possible to analyse huge amounts of climate data, such as those related to temperature, precipitation, atmospheric conditions, and more. This data's patterns and trends may be easily identified by machine learning algorithms, allowing for a greater comprehension of climate dynamics.
2. Enhanced Climate Modelling: AI is used to improve climate models, making them more complicated and better able to simulate intricate interactions within the Earth's climate system. This helps scientists evaluate the possible effects of climate change by enabling more precise projections of future climatic scenarios.
3. Prediction of Extreme Weather: AI systems can examine past meteorological data to find trends linked to catastrophic weather occurrences including hurricanes, droughts, and floods. Early warning systems cannot function without this predictive skill, which enables communities to efficiently plan for and respond to imminent calamities.
4. Ecosystem Monitoring: AI helps assess the condition of ecosystems that are impacted by climate change. This involves tracking changes in vegetation, biodiversity, and other ecological markers using satellite images and sensor data processing, which offers crucial insights for conservation efforts.
5. Tracking Carbon Emissions: AI-enhanced monitoring systems can monitor and examine carbon emissions from a variety of sources, including transportation and industry. This data is necessary for determining whether emission reduction goals have been met and for developing mitigation plans for the climate-changing effects of greenhouse gases.

6. Adaptation Planning: By assessing climate data and identifying regions vulnerable to changes, AI helps to build adaptation plans. Planning for infrastructure resilience, water resource management, and other actions to adapt to the changing climatic conditions are made easier with the use of this information.

7. Renewable Energy Optimisation: AI is used to optimise how renewable energy sources are integrated into the electrical system. Artificial intelligence (AI) contributes to improving the efficiency and dependability of sustainable energy systems by forecasting energy demand and maximising the production of renewable resources like solar and wind.

Ocean monitoring is made easier and more accurate with the use of AI technology. This includes monitoring ocean currents, sea surface temperatures, and marine biodiversity. These discoveries help us comprehend the part seas play in controlling the climate.

8. Early Climate Anomaly Detection: AI systems are able to spot anomalies or strange patterns in climate data that might point to upcoming climate trends or disasters. Early detection enables preventative actions and treatments to deal with possible problems.

9. Worldwide Cooperation: AI-enhanced climate monitoring systems promote worldwide cooperation by making it possible for academics and organisations to share data, models, and insights. Our ability as a society to solve the intricate and interwoven concerns of climate change is improved by this collaborative approach [8].

AI-enhanced climate monitoring systems provide a potent and revolutionary method for comprehending, reducing, and preparing for the effects of climate change. By utilising artificial intelligence to handle, analyse, and understand the huge quantity of data necessary for climate research and decision-making, these systems contribute to a more sustainable and resilient future.

1.3.2   Precision Agriculture and Resource Management

In addition to being essential for maximising crop output, the synergy between precision agriculture and resource management is also in line with more general sustainability and eco-friendly activities, notably in the context of combating climate change [9]. Precision agriculture, with its data-driven methodology, is essential for developing productive and sustainable agricultural practices. Precision agriculture turns traditional farming into a highly effective and environmentally beneficial activity by utilising technology like GPS, sensors, and artificial intelligence.

Data-driven decision-making, uses of variable rate technology (VRT), control of soil health, and precise irrigation techniques are all parts of resource management in precision agriculture. These techniques reduce resource waste and adverse environmental effects, improving the sustainability of agriculture. For instance, VRT enables farmers to apply inputs like insecticides and fertilisers at varying rates to accommodate the demands of different fields. By reducing overapplication, this accuracy improves crop quality while lowering environmental pollution.

By examining soil characteristics, nutrient concentrations, and pH, precision agriculture also emphasises managing soil health. Optimising nitrogen application minimises soil erosion, reduces nutrient runoff, and improves overall soil health, all of which help ensure long-term sustainability. Additionally, accurate irrigation techniques that are based on current sensor data and weather forecasts help to preserve water resources and use less energy.

1.3.2.1 AI in Climate Change Mitigation

AI is essential in combating climate change and promoting environmentally responsible farming, working in tandem with precision agriculture and resource management. Machine learning techniques are used by AI-enhanced climate monitoring systems to assess data from a variety of sources, such as satellites, sensors, and weather stations [10]. AI can give early warnings for extreme weather occurrences by spotting patterns and trends, helping with disaster planning.

AI-driven climate models improve the precision of climate projections, enabling us to comprehend climate dynamics and evaluate the potential effects of climate change. These models give us information about changes in temperature, precipitation, and weather patterns, enabling us to create proactive methods of reducing the effects of climate change.

The optimisation of renewable energy is another area where AI is useful. AI maintains a steady and sustainable power supply by forecasting energy demand and maximising the production of renewable energy sources like wind and solar electricity. Sustainability and environmental friendliness are further aligned with this shift to sustainable energy sources [11].

By maximising resource use, conserving water, reducing greenhouse gas emissions, and improving crop quality, this approach not only addresses climate change but also promotes a greener and more sustainable future in agriculture. The convergence of precision agriculture, resource management, and AI is a promising path to sustainable and eco-friendly farming practises.

1.3.3 Smart Energy Grids and Renewable Integration

The management of renewable energy sources will be transformed by the use of AI in energy networks, furthering the goal of environmentally benign power generation. By intelligently balancing electrical supply and demand, AI algorithms play a crucial part in improving the performance of energy networks and lowering reliance on traditional "dirty" energy sources [12]. Additionally, AI models play a critical role in forecasting the availability of solar and wind energy, permitting the smooth integration of renewable energy sources into current systems.

The search for sustainable and environmentally friendly energy generation has given a boost to renewable energy sources like solar and wind energy. However, the instability of the grid is hampered by their sporadic and unpredictable character. AI can provide creative methods to fully use renewable energy sources while ensuring grid dependability in this situation.

Incorporating information about the weather, the time of day, and previous consumption trends, AI-driven energy management systems are capable of forecasting electricity demand with astounding precision. In order to maintain a steady supply of electricity while maximising the usage of renewable resources, grids may modify supply in real-time thanks to these predictive capabilities.

Grid stability is a key factor in the integration of renewable energy. Variations in the amount of electricity produced from renewable sources can cause grid instability. AI algorithms continually monitor the state of the grid, quickly reacting to issues and assisting in maintaining ideal voltage and frequency levels. The resilience of the system and the prevention of power outages both depend on this degree of management.

Additionally, AI can estimate the availability of renewable energy sources. Grid operators can plan effectively for energy generation and distribution by using AI models that can predict wind and solar irradiance. These forecasts ensure that renewable energy is adequately utilised and help reduce energy waste and system congestion [13].

A game-changer for the switch to sustainable energy is the incorporation of AI in electricity systems. It lessens the usage of fossil fuels, lowers greenhouse gas emissions, and makes it possible to generate electricity sustainably. The role of AI in managing renewable energy is crucial as we work to battle climate change and build a cleaner future.

1.3.4   FOREST CONSERVATION AND WILDLIFE PROTECTION WITH AI

Two urgent worldwide issues are the preservation of forests and the safeguarding of threatened animal species. AI technologies are proving to be useful tools in this context for protecting our natural ecosystems. Drones with AI algorithms can be crucial in spotting and stopping illegal forestry operations by giving authorities fast and useful information [14]. Additionally, deforestation trends may be found via satellite imagery analysis using machine learning algorithms, which aids in conservation efforts and supports replanting programmes.

1.3.4.1   AI-Enhanced Forest Surveillance with Drones

Drones powered by AI have transformed forest protection and monitoring. These unmanned aerial vehicles may be used in isolated wooded regions to keep an eye out for unauthorised logging activities since they are fitted with cutting-edge image recognition and machine learning algorithms [15]. High-resolution photos and videos are taken by drones and are instantly analysed. Rapid detection of suspicious activity, such as unauthorised tree cutting, enables authorities to act quickly and stop future forest harm. This ongoing observation not only serves as a deterrent but also offers important proof in cases where illegal loggers are being sued.

1.3.4.2   Satellite Imagery and Deforestation Analysis

Massive volumes of satellite footage may be analysed by AI-powered machine learning algorithms to spot and track trends of deforestation. These algorithms look for alterations in forest cover, unpermitted logging, and intrusions into protected areas. For early intervention and the implementation of environmental legislation, such analysis is essential [16]. Additionally, by identifying appropriate places for restoration, it aids reforestation operations and informs conservation policies. By tracking the effects of deforestation on biodiversity and ecosystems, historical satellite data analysis aids in the development of evidence-based conservation strategies.

1.3.4.3 The Role of AI in Wildlife Protection

AI technology extends beyond forest monitoring to include animal conservation. The analysis of data from video traps and acoustic sensors using machine learning algorithms may be used to track the distribution of wildlife and identify poaching incidents. Assuring the protection of wildlife and discouraging poachers, AI-based surveillance systems in protected areas may automatically identify and inform authorities to the presence of endangered species or suspicious human activity [17].

In conclusion, AI technologies are revolutionising wildlife protection and forest conservation. They boost the efficiency and efficacy of conservation activities by supplying real-time monitoring and analysis capabilities. These technologies enable reforestation efforts and the preservation of crucial ecosystems in addition to helping to stop illicit activity.

1.3.5 CLIMATE CHANGE MODELLING AND PREDICTION WITH AI

Understanding, foreseeing, and mitigating the effects of climate change depend greatly on AI-driven climate modelling and prediction systems. To analyse past climate data, spot detailed trends, and simulate the future, these systems use machine learning techniques. This offers scientists and decision-makers priceless knowledge to create successful plans to combat climate change.

1.3.5.1 AI for Improved Climate Modelling

Climate data often contains complicated and nonlinear interactions, which machine learning algorithms can handle with ease. AI-driven systems may identify tiny patterns and correlations in huge datasets, in contrast to traditional methods. By capturing the dynamic character of Earth's climatic processes, this improved knowledge enables more precise climate modelling. For instance, research by [7] shows how satellite-based machine learning is applied to monitor and predict forest cover reduction, demonstrating the flexibility of AI in environmental research.

1.3.5.2 Accurate Forecasting of Climate Impacts

Based on past climate data, AI algorithms excel in making precise projections. These algorithms may model probable future situations by analysing trends and patterns and giving insights into the anticipated effects of climate change. Foreseeing changes in temperature, precipitation, sea levels, and extreme weather occurrences requires the ability to foresee the future. It supports the creation of adaptive policies by policymakers to safeguard ecosystems and communities. The need of exact classification techniques is highlighted [6]. This idea is relevant to climate modelling, as correct categorization is necessary for reliable predictions.

1.3.5.3 Informed Decision-Making for Climate Action

Decision-makers are empowered by evidence-based knowledge thanks to AI-driven climate models. These models assist policymakers in determining the efficiency of various mitigation and adaptation policies by simulating various climate scenarios. The useful information gleaned by AI analysis aids in the creation of educated policies and cross-border partnerships to tackle the problems associated with climate change. The research by [5] on the use of unmanned aerial vehicles (UAVs) in forestry emphasises the importance of cutting-edge technologies in environmental monitoring, which is consistent with the larger subject of utilising technology for insights connected to climate.

An innovative strategy for combating climate change is the use of AI in systems for climate modelling and prediction. These technologies provide precise forecasting, promote well-informed decision-making, and offer a thorough knowledge of climate dynamics. Artificial intelligence (AI) is a potent partner in the fight for a sustainable and resilient future as the global community deals with previously unheard-of environmental challenges [18].

1.3.6 Enhancing Disaster Prevention and Response Systems with AI

AI technologies have become crucial instruments for strengthening catastrophe prevention and response systems, greatly enhancing community sustainability and resilience [19]. These systems can analyse large datasets from several sources, such as weather predictions and social media updates, by utilising artificial intelligence. This analytical skill provides preventative efforts against climate change-related disasters and ensures quick and effective responses in disaster-affected areas [20].

1.3.6.1 Early Warning Systems for Climate-Related Disasters

Early warning systems powered by AI are essential for preventing disasters. Large volumes of data, such as satellite images, historical catastrophe records, and meteorological data, may be processed using machine learning algorithms. Whether it be a storm, flood, or wildfire, these algorithms find patterns and abnormalities that can signal the beginning of a calamity. AI can offer prompt alarms by analysing these patterns, giving populations vital time to escape or be ready for the imminent threat. The study of [8] emphasises the integration of AI in early warning systems, highlighting the value of cutting-edge technology in enhancing catastrophe preparedness.

1.3.6.2 Data Fusion and Situational Awareness

Real-time information and a thorough comprehension of the changing circumstances are essential for disaster response. AI is excellent at fusing data from several sources, including social media, satellite imaging, and on-the-ground sensors. Emergency responders' situational awareness is improved by AI systems by collecting and analysing this variety of data. They can then decide with confidence where to allocate resources, how to evacuate, and how to deploy. Research like that done by [8] emphasises the necessity of integrated information for successful disaster response and the role of AI in data fusion and situational awareness.

1.3.6.3 Resource Optimization and Predictive Modelling

When responding to disasters, AI helps to allocate resources as efficiently as possible. Based on past data and present conditions, machine learning algorithms can forecast the potential effect and spread of disasters. This predictive modelling aids emergency services in more effectively allocating resources, including manpower, medical supplies, and equipment. AI may also streamline logistics, ensuring that relief gets to afflicted areas on time and in a well-organized manner. Studies like those by [15], demonstrating the potential of cutting-edge technology in enhancing disaster response capabilities, are an example of the application of AI in resource optimisation.

As a result, the incorporation of AI technology into systems for disaster prevention and response marks a paradigm change in our capacity to safeguard populations against natural disasters. These technologies increase the effectiveness of resource utilisation during times of need while also enabling early alerts and better situational awareness. AI is a critical ally in creating resilient and sustainable communities as climate-related disasters increase in frequency and severity.

1.3.7 SUSTAINABLE TRANSPORTATION AND MOBILITY SOLUTIONS

1.3.7.1 Introduction to Sustainable Transportation

The environment, public health, and the standard of living in our communities are significantly impacted by the way we transport people and things. Climate change, air pollution, and traffic congestion are all greatly exacerbated by unsustainable transportation practises, which are characterised by a strong dependence on fossil fuels and ineffective systems [21]. The struggle against climate change and the pursuit of a cleaner future have made the switch to sustainable transport options imperative. This introduction lays the groundwork for examining the tenets, approaches, and technological advancements that support sustainable transportation and their critical contribution to climate change mitigation.

1.3.7.1.1 Definition and Goals of Sustainable Transportation

We go into the fundamental ideas of sustainable transportation in this section. A collection of practices and regulations known as "sustainable transportation" strives to lessen the detrimental effects of transportation on the environment and society while maintaining its economic viability and accessibility. We look at sustainable transportation's objectives, which include lowering greenhouse gas emissions, encouraging public transportation and active transportation, boosting energy efficiency, and raising the standard of living in both urban and rural areas. The explicit definition of sustainable transportation and its many goals is established in this subtopic.

1.3.7.1.2 Strategies for Sustainable Transportation

The strategies cover a wide range of topics, including increasing active mobility, supporting public transportation, lowering the usage of private automobiles, developing electric and driverless vehicles, and streamlining logistics and freight transportation. We examine the crucial roles that technology, urban design, and cutting-edge governmental initiatives play in these tactics. The subtopic offers a thorough analysis of the actions undertaken in realising the objectives of sustainable transportation.

1.3.7.1.3 The Role of This Book in Exploring Sustainable Mobility Solutions

This describes the goals of the book, which are to educate and motivate readers about the potential and difficulties of sustainable transportation options. Researchers, decision-makers, and anybody else interested in learning more about how important transport is to reducing climate change may find this book to be a useful resource. This subtopic provides a road map for the chapters and issues that will be investigated, showing how each chapter advances sustainable transportation practises in general.

1.3.8 Revolutionising Waste Management through AI

Artificial intelligence (AI) technology integration in waste management shows tremendous promise for improved trash sorting efficiency, recycling techniques, and overall effectiveness [22]. Classifying waste products, streamlining garbage collection routes, and minimising environmental effects are all made possible by AI-driven systems that make use of machine learning algorithms and intelligent sensors.

1.3.8.1 Intelligent Waste Sorting with Machine Learning

Machine learning techniques are used by AI-powered garbage sorting systems to effectively classify various waste items. These algorithms can correctly detect and separate recyclables from non-recyclables by analysing visual data from cameras or sensors. By reducing contamination and encouraging the recovery of valuable materials, this improves the effectiveness of recycling operations. The use of machine learning in garbage sorting is explored in research by [18], with a focus on the potential for AI to increase recycling rates and lessen the environmental effect of waste disposal.

1.3.8.2 Smart Sensors for Optimized Waste Collection

To track garbage levels and improve collection routes, waste bins are equipped with AI-powered sensors. These sensors monitor the degree of bin fill using real-time data, allowing waste management services to arrange pickups more effectively. AI helps cut down on the pollutants produced by garbage collection vehicles by preventing needless collections of bins that are only partly filled. The study by [12] digs into the installation of smart sensors in garbage cans, showing their importance in reducing the amount of fuel used and environmental pollution caused by ineffective waste collecting methods.

1.3.8.3 Predictive Analytics for Waste Management

Predictive analytics, which forecasts future trash generation patterns based on past data and present trends, is made possible by AI technology in waste management. With this proactive strategy, governments and trash management businesses may plan for greater recycling capacity, allocate resources more efficiently, and carry out targeted educational efforts to minimise particular types of garbage. In their study, [17] highlights the potential for data-driven decision-making in the optimisation of waste management methods by discussing the incorporation of AI in predictive analytics for trash management.

By introducing intelligent sorting, streamlining garbage collection routes, and facilitating data-driven decision-making, AI technologies are altering waste management and recycling practises. These developments not only enhance the rates of resource recovery and recycling but also lessen the environmental effects of ineffective waste management techniques.

1.3.9 ADVANCING SUSTAINABLE BUILDING PRACTICES WITH AI

An innovative strategy for maximising energy usage, lowering emissions, and improving overall efficiency is the integration of artificial intelligence (AI) into green building and sustainable design initiatives [23]. Artificial intelligence (AI) supports architects and designers in modelling energy efficiency and material consumption scenarios for the construction of environmentally friendly structures. AI-driven smart building management systems are vital for automating and optimising energy use.

1.3.9.1 AI-Driven Smart Building Management Systems

The optimisation of energy use depends heavily on AI algorithms in smart building management systems. To decide when to light, heat, and cool a space, these systems employ real-time data from sensors. Artificial intelligence (AI) makes sure that buildings run as efficiently as possible by recognising patterns of usage and adapting quickly to changes in the environment. The usefulness of AI in smart buildings is shown by research by [14], highlighting its significance in attaining energy efficiency and sustainability goals. In addition to lowering energy costs, smart building solutions help lessen the total environmental effect of building operations.

1.3.9.2 AI-Assisted Design for Green Buildings

By modelling and evaluating numerous design situations, AI technologies help architects and designers create green structures. The energy efficiency of various architectural components may be assessed, sustainable materials can be suggested, and layouts for natural lighting can be optimised using machine learning techniques. Informed design decisions may then be made by architects to improve the building's overall sustainability. The study by [17] investigates how AI is incorporated into architectural design processes and illustrates how AI contributes to creative and sustainable building design.

1.3.9.3 Predictive Maintenance for Sustainable Infrastructure

Predictive maintenance techniques for environmentally friendly infrastructure use AI. Artificial intelligence (AI) can anticipate possible equipment breakdowns and inefficiencies by analysing data from sensors integrated into building systems. In addition to cutting down on downtime and maintenance expenses, this proactive strategy makes sure that building systems are running at their peak efficiency, which promotes long-term sustainability. The study by [16] examines the use of AI in preventative maintenance for sustainable infrastructure, with special emphasis on how it helps to guarantee the dependability and effectiveness of building systems.

AI is a potent ally in promoting sustainable design and green building techniques. The efficiency and environmental friendliness of buildings are improved by AI technologies, which help with everything from everyday operations to design. The incorporation of AI is a critical factor in producing energy-efficient, ecologically responsible, and creative construction solutions as the need for sustainable infrastructure increases.

1.3.10　AI Empowering Environmental Education and Awareness

In the field of environmental education, artificial intelligence (AI) is a potent tool that provides individualised knowledge, encourages interactive interaction, and promotes eco-friendly behaviours. Intelligent chatbots, virtual assistants, and visualisations produced by AI all aid significantly in teaching people about sustainable practices [24].

1.3.10.1　Personalized Guidance through AI Chatbots

Chatbots using AI act as individualised tour guides for environmental education. These smart technologies may initiate dialogues with users while offering personalised advice, sustainable practice ideas, and recommendations. Chatbots' flexibility makes sure that users get information that is pertinent to their situation and interests. The potential of AI chatbots in environmental education is explored by research by Mader et al. (2020), with a focus on their function in delivering user-centric and customised material to encourage sustainable behaviour.

1.3.10.2　Virtual Assistants for Eco-Friendly Habits

AI-powered virtual assistants who provide timely advice and recommendations help people develop environmentally beneficial behaviours. These aides can help users make sustainable decisions in their daily lives, such as trash reduction tactics and energy-efficient practises. The study by [18] demonstrates the interactive and supporting role that AI plays in encouraging eco-friendly habits by emphasising the usage of virtual assistants.

1.3.10.3　AI-Generated Visualizations for Engagement

Audience engagement and the communication of difficult environmental ideas are greatly enhanced by AI-generated visualisations and simulations. By streamlining information, these visualisations can make it more comprehensible and appealing. AI-generated pictures help people grasp complex concepts and inspire a commitment to live sustainably, whether they are used to highlight the effects of climate change or to illustrate the advantages of sustainable practises. In their study on the use of visualisations in environmental communication, [20] emphasise the value of aesthetically attractive and educational material in spreading environmental messages.

AI makes a substantial contribution to environmental education and awareness by providing individualised help, interactive features, and captivating visualisations. The role of AI in educating people with information and fostering eco-friendly practises is vital for creating a more environmentally conscious global community as the need for sustainability becomes more pressing.

1.4　LEVERAGING AI FOR A SUSTAINABLE FUTURE

There are countless opportunities to address the urgent issues caused by climate change and to promote the development of a sustainable and environmentally friendly environment as a result of the integration of artificial intelligence (AI) into many areas of society [25]. Humanity may greatly contribute to the mitigation of climate change and the advancement of a greener and more resilient future by utilising AI's skills in monitoring the environment, precision agriculture, integrating renewable energy, and other disciplines. The adaptive intelligence and comprehensive understanding of AI are set to play a key role in defining the sustainable solutions of the future as academics and policymakers explore the technology's broad possibilities.

Artificial intelligence (AI) is a revolutionary force in our search for sustainability in a time of growing environmental concerns. AI-driven climate monitoring systems offer priceless insights into climate dynamics, allowing us to foresee and effectively address climate-related concerns. With the use of AI technology, precision agriculture improves agricultural yields, reduces environmental impact, and better manages resources. The use of AI in intelligent energy networks promotes the production of green energy while balancing energy supply and demand and minimising reliance on harmful sources. AI technology also facilitates effective trash management and improves recycling procedures, thus reducing environmental footprints.

In order to fully use AI for a sustainable future, researchers, politicians, and engineers must work together. We can establish plans to stop global warming, switch to renewable energy sources, and create eco-friendly urban infrastructure through improving AI solutions. AI-driven technologies are changing the way we think about protecting the environment by enabling us to take both individual and group action by forming wise judgements and implementing sustainable practises.

It is essential to continue to be committed to ethical and responsible AI development as AI's role in sustainability changes over time. In order to keep AI as a potent force for good change, it is crucial to strike a balance between technical development and environmental protection.

As a whole, the application of AI to the cause of sustainability is more than just a technological development; it symbolises a common desire to protect the environment and ensure a better, greener, and more sustainable future for future generations.

1.5 APPLICATIONS OF AI IN ENVIRONMENTAL SUSTAINABILITY

The use of AI in environmental sustainability has many different applications. Its basic responsibilities include limiting emissions and maximising energy use. Buildings, businesses, and transportation systems may monitor and manage their energy use with the help of AI-powered algorithms, which will significantly reduce energy waste and greenhouse gas emissions. For instance, smart grids use AI to effectively manage power supply and demand, and predictive maintenance reduces energy losses in industrial equipment [26].

Agriculture is another area where AI is crucial. The world's expanding population puts increasing strain on resource use and food supply. AI-driven precision agriculture maximises crop management, resource allocation, and pest control by using sensors and data analytics, increasing agricultural yields while reducing environmental impact [27].

Additionally, AI helps conservation efforts by keeping an eye on and safeguarding habitats and species. Large datasets may be analysed by machine learning algorithms, which can then be used to track and protect endangered species, safeguard biodiversity, and stop unlawful poaching [28].

Another essential component of sustainable environmental practises is the use of renewable energy sources. AI helps increase the effectiveness of renewable energy sources like wind and solar electricity. AI aids in the integration of clean energy sources into the current infrastructure while maintaining a reliable and sustainable power supply through energy storage optimisation, smart grids, and predictive maintenance [29].

AI also helps with disaster management and climate forecast. In order to improve climate forecasts and provide early warnings for catastrophic weather occurrences like hurricanes, floods, and droughts, machine learning algorithms can examine past meteorological data. For resilience and catastrophe preparedness, this ability is essential.

In conclusion, it is becoming increasingly clear that AI has the adaptability and capacity to handle climate concerns and advance environmental sustainability. AI is playing a key role in determining a sustainable and environmentally friendly future due to its capacity to handle massive quantities of data, make predictions, and optimise resource consumption.

1.6 CASE STUDIES AND SUCCESS STORIES FOR CLIMATE CHANGE

Concrete examples of the efficiency of AI-driven solutions in combating climate change and accomplishing sustainability goals are provided in the section on "Case Studies and Success Stories" [29]. It outlines some of the prominent success stories and provides instances of how AI has really been used to improve the environment.

Real-World Examples of AI-Driven Solutions:

This section focuses on concrete examples of how AI technology has been used to address climate change. Case examples show how AI-driven climate monitoring systems have revolutionised environmental data collecting and analysis, enabling more precise forecasts and better-informed decision-making. Other examples might include AI-integrated renewable energy management, showing how AI algorithms can effectively integrate solar and wind energy into current power systems while balancing supply and demand [29]. The section may also describe how AI might be used to make big cities' public transport systems better, lessen traffic, and promote sustainable mobility options.

Demonstrations of AI's Role in Achieving Sustainability Goals:

AI's revolutionary potential in achieving sustainability goals is highlighted in this section. It explains how drones with AI algorithms have been crucial in identifying illicit logging activities and assisting in the conservation of endangered species. It also discusses AI's contributions to forest conservation and animal preservation. The section may also include AI-driven climate change prediction models that show how machine learning algorithms examine past climate data to simulate possible futures and offer insightful information to decision-makers. Additionally, the crucial role played by AI in systems for disaster prevention and response is highlighted, highlighting the way in which AI-driven systems examine enormous datasets to enable pre-emptive measures and effective reactions in catastrophes connected to climate change [28].

The success stories and case studies serve as motivating illustrations of AI's potential to combat climate change and advance environmental sustainability. These real-world examples offer a clear picture of the practical uses and beneficial results of AI-driven solutions, which encourages ongoing research and deployment of these technologies to address climate concerns and build a more sustainable future.

1.7 FUTURE TRENDS AND OPPORTUNITIES IN AI FOR ENVIRONMENTAL SUSTAINABILITY WITH A FOCUS ON CLIMATE CHANGE

1.7.1 AI-DRIVEN CLIMATE RESILIENCE

Improving climate resilience will be essential to AI's role in environmental sustainability, especially considering climate change. The ability of AI models to forecast extreme weather, sea level rise, and other climate-related dangers will improve, enabling people to better adapt and prepare.

There are several possibilities for creating AI systems that offer real-time climate data, early warnings, and adaptive solutions to lessen the effects of climate-related calamities and create communities that are robust to climate change.

1.7.2 AI-ENABLED CLIMATE POLICY SUPPORT

AI will be essential in assisting decision-makers in developing sensible climate policies. Future AI systems will offer comprehensive climate modelling and scenario analysis, assisting governments in establishing challenging goals, monitoring progress, and improving policies for sustainability and emissions reduction.

Fostering global collaboration, allowing data-driven climate accords, and encouraging the adoption of evidence-based climate policy are some of the potential uses in this field.

1.7.3 AI FOR CARBON SEQUESTRATION AND REMOVAL TECHNOLOGIES

Future AI-driven solutions will concentrate on improving carbon capture and removal processes. Direct air capture (DAC) technologies and carbon capture and storage (CCS) systems can both benefit from AI-enhanced design and operation, thereby expediting the development of carbon-neutral solutions.

The creation of AI-assisted CCS and DAC technology presents opportunities to dramatically advance carbon reduction efforts and aid in the transition of the world to a low-carbon and sustainable economy.

Addressing the pressing issues of a warming planet requires using these trends and potential in AI for environmental sustainability, with an emphasis on climate change in particular. AI is positioned to drive breakthroughs that will greatly help to the battle against climate change and the building of a more sustainable and eco-friendly society by boosting climate resilience, promoting effective climate policies, and expanding carbon sequestration technology.

1.8　CONCLUSION

In the ongoing fight against climate change, the incorporation of artificial intelligence (AI) into different facets of society is a revolutionary force with limitless potential to build sustainable and environmentally friendly habitats. These AI-driven solutions cover a wide range of crucial fields, such as real-time climate monitoring, resource management, precision farming, smart grid optimisation, and renewable energy integration. AI also plays a crucial role in preserving important ecosystems, safeguarding threatened species, and directing our efforts to mitigate climate change through precise modelling and prediction. While sustainable transportation and mobility solutions lessen carbon footprints and increase efficiency, disaster preventive and response systems guarantee community safety and resiliency. The construction sector is revolutionised by AI's adoption of green building and sustainable design principles, which redefines our approach to waste disposal while promoting recycling and resource conservation. Additionally, it involves people and communities in campaigns to raise environmental consciousness.

These AI-driven solutions represent not only technology development but also a shared commitment to protecting the environment. They provide a wide range of options for halting climate change, lowering emissions, and building a greener, more sustainable future. The path to sustainability through AI is not without its difficulties, though. As AI continues to play a crucial role at the nexus of technological innovation and environmental stewardship, ethical concerns, responsible AI development, and international cooperation are vital elements of this endeavour. In conclusion, the incorporation of AI offers solutions that can improve our world and offer a ray of hope. A journey marked by optimism, innovation, and the unwavering belief that we can create a world where environmental sustainability is not just a goal but a reality; embracing AI technologies is not just about addressing the challenges of today but also about defining the sustainable solutions of tomorrow.

REFERENCES

1. Khan, M. A. U., & Malik, R. K. (2020). Precision agriculture technologies for crop management. In Recent Trends and Future Technology in Agriculture and Food Engineering (pp. 241–266). Springer.
2. Rezgui, Y., Raja, B. Y., & Iqbal, N. (2021). Artificial Intelligence for Energy Efficiency in Smart Buildings: A Comprehensive Review. Energy and AI, 3, 100059.
3. www.irena.org/publications/2019/Jun/AI-for-Good-Artificial-Intelligence-in-Renewable-Energy

4. "AI for Good: Artificial Intelligence in Renewable Energy"—IRENA (International Renewable Energy Agency). (2019). www.irena.org/publications/2019/Jun/AI-for-Good-Artificial-Intelligence-in-Renewable-Energy

5. Guo, Y., et al. (2020). Application of Unmanned Aerial Vehicles (UAVs) in Forestry. Remote Sensing, 12(13), 2078.

6. Lu, D., & Weng, Q. (2007). A Survey of Image Classification Methods and Techniques for Improving Classification Performance. International Journal of Remote Sensing, 28(5), 823–870.

7. Rao, N. D., & Niles, M. T. (2021). Satellite-based Machine Learning for Monitoring Forest Cover Loss in Sumatra and Kalimantan, Indonesia. PLoS ONE, 16(7), e0253436.

8. Lee, J., et al. (2020). A Review of Artificial Intelligence for Disaster Management. Disaster Advances, 13(3), 48–53.

9. Chen, S., et al. (2018). A Survey of Artificial Intelligence in Disaster Management: Opportunities, Challenges, and Future Directions. ACM Computing Surveys, 51(1), 7.

10. Wang, Y., et al. (2021). Machine Learning for Waste Sorting and Recycling: A Comprehensive Review. Waste Management, 119, 37–48.

11. Riziotis, V., et al. (2019). Development and Evaluation of an Ultrasonic Sensor-based Waste Bin Level Monitoring System. Resources, Conservation and Recycling, 141, 83–94.

12. Makridakis, S., et al. (2018). The Forthcoming Artificial Intelligence (AI) Revolution: Its Impact on Society and Firms. Futures, 90, 46–60

13. Ma, Z., et al. (2020). Artificial Intelligence in Smart Buildings: A Comprehensive Survey. IEEE Transactions on Industrial Informatics, 16(3), 2157–2164.

14. Cheng, Y., et al. (2018). Deep Convolutional Neural Network-based Regression Approach for Estimating the Energy Performance of Office Buildings. Energy and Buildings, 177, 235–246.

15. Zhang, Q., et al. (2019). Predictive Maintenance of Sustainable Infrastructure Using Artificial Intelligence. Journal of Building Performance, 10(1), 51–64.

16. Mader, C., et al. (2020). Chatbot Acceptance in Environmental Education: The Role of User Factors and Chatbot Attributes. Frontiers in Artificial Intelligence, 3, 40.

17. Wang, X., et al. (2018). Development of a Virtual Assistant for Promoting Environmentally Friendly Behaviors. Sustainability, 10(3), 580.

18. Ferreira, N., et al. (2014). Visual analytics for making sense of large-scale environmental data. In EuroVis (Short Papers) (pp. 61–65). Springer.

19. Chen, L., Chen, Z., Zhang, Y., et al. (2023). Artificial Intelligence-Based Solutions for Climate Change: A Review. Environmental Chemistry Letters, 21, 2525–2557. https://doi.org/10.1007/s10311-023-01617-y.

20. IPCC Special Report on Global Warming of 1.5°C. (2018). Intergovernmental Panel on Climate Change. IPCC.

21. Gandomi, A., & Haider, M. (2015). Beyond the Hype: Big Data Concepts, Methods, and Analytics. International Journal of Information Management, 35(2), 137–144.

22. Sharma, H. B., et al. (2019). Artificial intelligence (AI) in healthcare and research. Proceedings of the 2019 3rd International Conference on Cloud and Big Data Computing (CBDCom). CBC.

23. Bonhomme, J., et al. (2020). A Review of Machine Learning Applications in Environmental Sciences. Progress in Physical Geography, 44(6), 697–730.

24. Caro, D., & Ali, A. (2019). Big Data and Artificial Intelligence as a New Paradigm of Scientific Research. Future Internet, 11(8), 175.

25. Bala, A., & Raj, R. G. (2019). Renewable energy applications using AI techniques. In Proceedings of the 2019 International Conference on Data Science and Machine Learning. IEEE.

26. Colley, R. T., et al. (2018). Deep Learning for Multi-modal Spectral Data Analysis: Applications in UV-visible and Infrared Spectroscopy. Analytica Chimica Acta, 1020, 1–10.
27. Khatami, A., et al. (2019). Machine Learning in Geosciences and Remote Sensing. Geoscience Frontiers, 10(1), 3–10.
28. Hussain, M., et al. (2020). Deep Learning-based Big Data Analytics for Smart Cities. Future Generation Computer Systems, 103, 534–546.
29. Williams, G. R., et al. (2020). Building Smart Cities: Emerging Trends and Challenges for Large-scale Urban IoT. IEEE Internet of Things Journal, 7(3), 2484–2505.

2 Enhancing Climate Change Prediction and Risk Assessment with Deep Learning
Architectural Approaches and Data Challenges

Deepak Gupta, Pradumn Kumar,
Man Mohan Shukla, and Satyasundara Mahapatra

2.1 INTRODUCTION

Climate refers to the collective average of weather conditions over a specific period of time. Average temperature, precipitation, humidity, and wind patterns are encompassed within this category. A lasting shift in the enduring weather patterns that have come to typify Earth's local, regional, and global climates characterizes climate change. The primary cause of climate change is human activities, primarily the burning of fossil fuels, leading to elevated levels of heat-trapping greenhouse gases in the Earth's atmosphere. These gases act as a blanket enveloping the Earth, capturing the sun's warmth and resulting in higher temperatures. The impact of climate change is already evident worldwide, including rising sea levels, glaciers melting, heightened instances of extreme weather events, shifts in agricultural production, and a reduction in biodiversity. Climate change poses a significant threat to both the planet and its inhabitants. It is crucial to take steps to reduce greenhouse gas emissions and mitigate the effects of climate change. Efforts to address climate change involve decreasing greenhouse gas emissions [1] by transitioning to renewable sources of energy (solar, wind, hydroelectric, etc.), increasing energy efficiency, reforestation, and adopting environmentally friendly technologies and sustainable land use practices. International pacts, including the Paris Agreement [2], strive to unite nations in their efforts to curtail global warming and reduce its impact. The Agreement establishes benchmarks for constraining the elevation of global temperatures, aiming to keep the rise well below 2 degrees Celsius compared to pre-Industrial levels and actively working towards limiting it to 1.5 degrees Celsius. Researchers worldwide are employing sophisticated deep learning techniques to evaluate the ramifications

DOI: 10.1201/9781003452393-2

of climate change. Researchers at UC Berkeley are making predictions about future sea level rise in the San Francisco Bay Area using deep learning [3]. The model is educated with past observations of tides, sea levels, and other variables. It can foretell future sea level rise as far out as half a century. Satellite photos and other data train the model to detect and track floods, droughts, and wildfires. Communities need this data to prepare for and react to catastrophes. The National Oceanic and Atmospheric Administration (NOAA) is modelling the economic implications of climate change using deep learning [4]. This model uses extensive data on meteorological trends, economic indicators, and other variables to predict the consequences of climate change on agriculture, energy, and transportation. The model then informs preventive and reactive climate change solutions.

2.2   DEEP LEARNING ARCHITECTURE

Deep learning is a subfield of machine learning that trains computer networks to solve problems like biological brain circuits. In recent years, there has been a lot of focus and development on its ability to solve complex data problems in a variety of fields. The foundation of deep learning models lies in neural networks, which comprise interconnected nodes, often referred to as "neurons" [5]. Each neuron processes inputs, performs mathematical operations on them, and generates corresponding outputs. The architecture of a neural network takes shape through layers of interconnected neurons. The term "deep" in deep learning denotes the incorporation of multiple layers within a neural network. This depth empowers the network to inherently grasp hierarchical data representations, enabling the discernment of intricate patterns and features. A pivotal advantage of deep learning lies in its autonomous acquisition of relevant features directly from raw data. In contrast, traditional machine learning necessitates the manual extraction of pertinent features. Exposure to large amounts of labeled data is essential for the development of deep learning models. Throughout the training process, the model adjusts its internal parameters via the mechanism of backpropagation. Fine-tuning the model's parameters to reduce error requires computing the gradient of the model's performance with respect to the parameters. A neural network's neurons use an activation function on their inputs, which introduces non-linearity to the system. This characteristic endows the network with the capability to effectively model complex relationships within the data. Diverse types of layers are employed within deep learning architectures, encompassing:

Input Layer: This layer accepts raw data and transmits it to the subsequent layer.

Hidden Layers: These intermediate layers positioned between the input and output layers serve the purpose of extracting features and establishing representations.

Output Layer: Responsible for generating the ultimate prediction or output of the model.

Training and evaluating deep learning models for climate change prediction and risk assessment requires a systematic approach that combines domain expertise, data preparation, model selection, and thorough evaluation. Here's a step-by-step guide:

a. Data Collection and Preprocessing:

- Gather relevant climate data, including historical records, satellite imagery, and model outputs.
- Thoroughly clean and preprocess the data, addressing any missing values and applying scaling and normalization techniques to variables for consistency.
- Split the data into training, validation, and test sets.

b. Model Selection:

- Choose an appropriate deep learning architecture based on the nature of the data and the prediction task (CNNs for images, RNNs/LSTMs for time series, etc.). Consider factors like model complexity, interpretability, and computational resources.

c. Model Design and Configuration:

- Construct the framework of the selected model, outlining the layer count, units, activation functions, and any other relevant hyperparameters. Infuse domain knowledge to influence model's arrangement and the extraction of features, culminating in a purpose-driven configuration.

d. Data Augmentation (Optional):

- Implement data augmentation methodologies to artificially expand the variety within the training data, particularly in the context of tasks involving images.

e. Training:

- Train the model using suitable optimization algorithms (e.g., Adam, RMSProp) and loss functions (e.g., Mean Squared Error, Cross-Entropy) with the training data. Keep a close watch on the training advancement by assessing metrics on the validation set, encompassing loss and accuracy.

f. Hyperparameter Tuning:

- Fine-tune hyperparameters (learning rate, batch size, etc.) to optimize the model's performance on the validation set.

g. Regularization Techniques (Optional):

- Apply regularization techniques like dropout, batch normalization, and weight decay to prevent overfitting.

h. Evaluation Metrics:

- Choose appropriate evaluation metrics based on the prediction task (e.g., Mean Absolute Error, Root Mean Squared Error, Precision, and recall). Additionally, consider domain-specific metrics if they are accessible and useful.

i. Model Evaluation:

- Determine the trained model's generalization ability by evaluating its performance on a test set. Scrutinize the model's predictions in relation to the actual data, encompassing quantitative metrics as well as visual representations.

j. Sensitivity Analysis (Optional):

- Conduct sensitivity analyses to understand how changes in input data affect model predictions and risk assessments.

k. Model Deployment (Optional):

- If applicable, deploy the trained model for real-time predictions or risk assessments.

Climate scientists, data scientists, and domain experts must work together throughout the process to ensure reliable modelling, insightful interpretation, and actionable outcomes.

2.3 DEEP LEARNING MODELS USED FOR CLIMATE CHANGE PREDICTION AND RISK ASSESSMENT

Various types of deep learning models can be applied to climate change prediction and risk assessment, each tailored to specific tasks and data characteristics. Some of the deep learning models commonly used in this context are:

a. **Convolutional Neural Networks (CNNs):** A CNN is a form of artificial neural network that replaces general matrix multiplication with a mathematical operation known as convolution in at least one of its layers. Multiple layers make up CNNs, with the convolutional layer, pooling layer, and fully connected layer being the most prominent. The convolutional layer accepts an input image and applies a convolution operation to extract features. The input image is slid across a filter, and the dot product between the filter and the image at each position is calculated. In the end, we get a feature map, which is a 2D array of numbers that stands in for the extracted features. It is common practice to use pooling layers to reduce the feature maps' dimensionality. To do this, either the largest value within a constrained area of the feature map is chosen, or the average value of that area is determined. This makes the network more resistant to noise and decreases the number of parameters it needs. Input images are categorized using fully connected

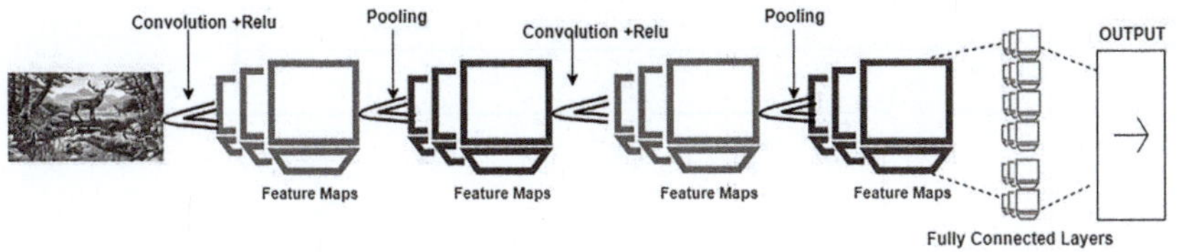

FIGURE 2.1 Convolutional neural network.

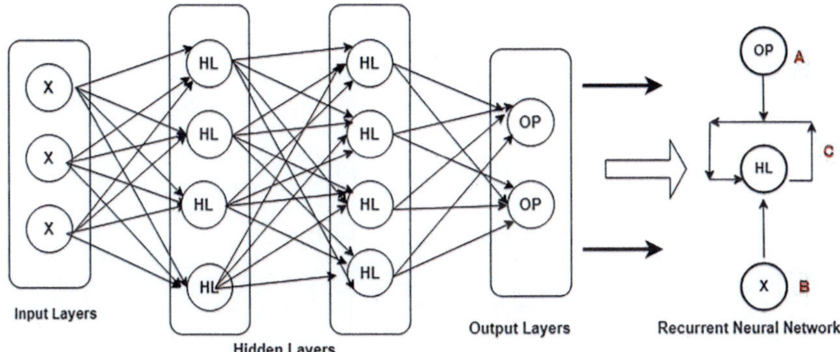

FIGURE 2.2 Recurrent neural network.

layers. They use a regular neural network to classify the images based on the feature maps produced by the convolutional layers. In particular, CNNs excel at processing satellite images and remote sensing data that pertain to climate. Cloud cover pattern recognition, sea ice monitoring, and land use change analysis are all possible applications [6].

b. **Recurrent Neural Networks (RNNs):** One type of Artificial Neural Network, called a "recurrent neural network" (RNN), is particularly well-suited to processing sequential input. Given the sequential structure of climate data, they offer hope as a method for predicting climate change. Since RNNs excel at processing sequential data with temporal relationships, they are well-suited for use with time series climate data. It's possible to use them for short-term weather forecasting and to record temporal trends in climatic variables like temperature swings [7].

c. **Long Short-Term Memory (LSTM) Networks:** The recurrent neural network (RNN) subclass known as LSTM networks excels at processing lengthy data sequences. The input history of an LSTM network is stored in a memory cell. Since longer-term trends and cyclic patterns, such as annual temperature changes, are essential for anticipating climate change, this feature of the network is crucial. The capacity of LSTMs to handle distant relationships and solve the vanishing gradient problem has significantly improved [8].

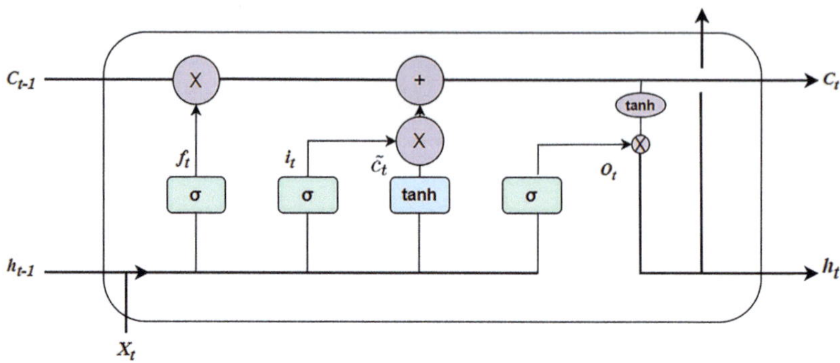

FIGURE 2.3 LSTM architecture.

$$i_t = \sigma(w_i[h_{t-1}, x_t] + b_i) \tag{1}$$

$$f_t = \sigma(w_i[h_{t-1}, x_t] + b_f) \tag{2}$$

$$o_t = \sigma(w_o[h_{t-1}, x_t] + b_o) \tag{3}$$

Here:

> i_t: input gate, f_t: forget gate, o_t: output gate, σ sigmoid function, w_x: weight for respective gate (x) neuron, h_{t-1}: output of previous lstm block (at timestamp t - 1), x_t: input at current timestamp, b_x: biases for respective gates (x).

Here, Equation 1 represents the input gate, which serves the purpose of determining the specific new information that will be stored inside the cell state. Equation 2 pertains to the forget gate, which serves the purpose of determining the information that should be discarded from the cell state. Equation 3 represents the output gate, which serves the purpose of generating the activation for the final output of the LSTM block at a certain timestamp, denoted as t.

> d. **Gated Recurrent Units (GRUs):** Like LSTMs, GRUs are a form of RNN, although they rely on fewer parameters. Because of this, they may be trained more quickly and with less sensitivity to the hyperparameters used. To regulate data entering and leaving the network, GRUs employ a gating mechanism. This lets the network zero in on the most pertinent data while passing over the rest. GRUs can detect temporal connections in sequential data. In particular, GRUs excel at time series forecasting. They can study the historical relationships between climate variables and predict the future with high precision [9].
>
> e. **Autoencoders:** Autoencoders belong to the category of artificial neural networks employed for acquiring latent data representations. This means that they can learn to represent data in a lower dimensional space, while

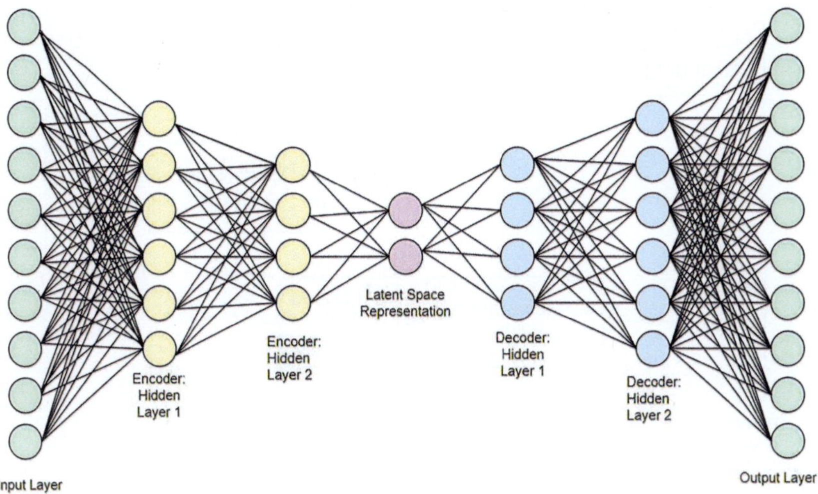

FIGURE 2.4 Autoencoder architecture.

still preserving the most important information. Autoencoders are unsupervised models used for dimensionality reduction and feature learning. They can help extract meaningful representations from high-dimensional climate data, aiding in data preprocessing and feature extraction [10].

f. Generative Adversarial Networks (GANs): Generative adversarial networks (GANs) represent a type of artificial neural network that holds the capability to produce lifelike data. This characteristic renders them a promising instrument for forecasting climate change, as they can produce synthetic climate data suitable for training other models. GANs operate by engaging two neural networks in a contest. The initial network, referred to as the generator, assumes the role of crafting new data. Conversely, the other network, termed the discriminator, shoulders the responsibility of distinguishing genuine data from counterfeit data. The generator persistently endeavors to deceive the discriminator, who, in turn, relentlessly strives to unveil the generator's ruse. This dynamic of competition and cooperation helps the generator learn how to produce more credible information. As time goes on and the generator gets better, the discriminator gets better at spotting fabricated data. This engenders a feedback loop that propels the enhancement of both networks. It also holds potential in fabricating climate change scenarios for the purpose of risk assessment [11].

g. Transformer Models: Transformer models constitute a class of artificial neural networks uniquely crafted for handling sequential data. This distinctive attribute positions them as a propitious instrument for forecasting climate change, given the inherent sequential structure of climate data. Transformer models operate by leveraging attention mechanisms to concentrate on the salient details within the sequence. This approach empowers the

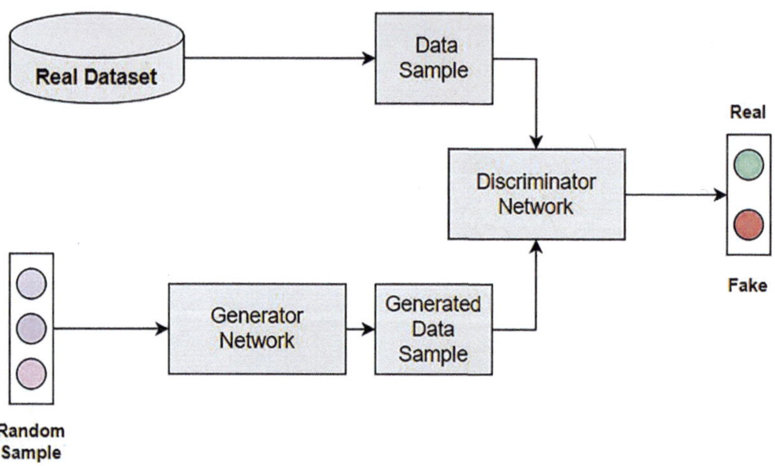

FIGURE 2.5 GAN architecture.

network to grasp extensive relationships within the data, a critical aspect for accurate climate change prediction [12].

h. Graph Neural Networks (GNNs): A graph neural network (GNN) is a type of artificial neural network that is specifically designed to process data that is represented as a graph. GNNs work by propagating information through the graph, using a process called message passing. This allows the network to learn the relationships between different entities in the graph, which is important for predicting climate change. GNNs are suitable for data with complex relationships, such as climate networks depicting interactions between different climate variables. They can be used to model climate data as a graph and analyze interdependencies between variables [13].

i. Ensemble Models: Ensemble models encompass a category of machine learning frameworks that amalgamate the forecasts generated by numerous models, leading to heightened accuracy. This strategy holds substantial potential in the context of climate change prediction, as it aids in rectifying the idiosyncrasies and inaccuracies inherent to singular models. A plethora of diverse ensemble techniques are available for employment in climate change prediction [8]. Among the most prevalent approaches are:

- **Bagging:** Bagging involves training multiple models on resampled subsets of the training data, employing a process known as bootstrapping. This approach serves to diminish the models' variance and elevate their overall accuracy.

- **Boosting:** Boosting, another approach, entails training multiple models in a sequential manner, wherein each subsequent model seeks to rectify the errors of its predecessors. This iterative correction fosters an augmentation of the aggregate models' accuracy.

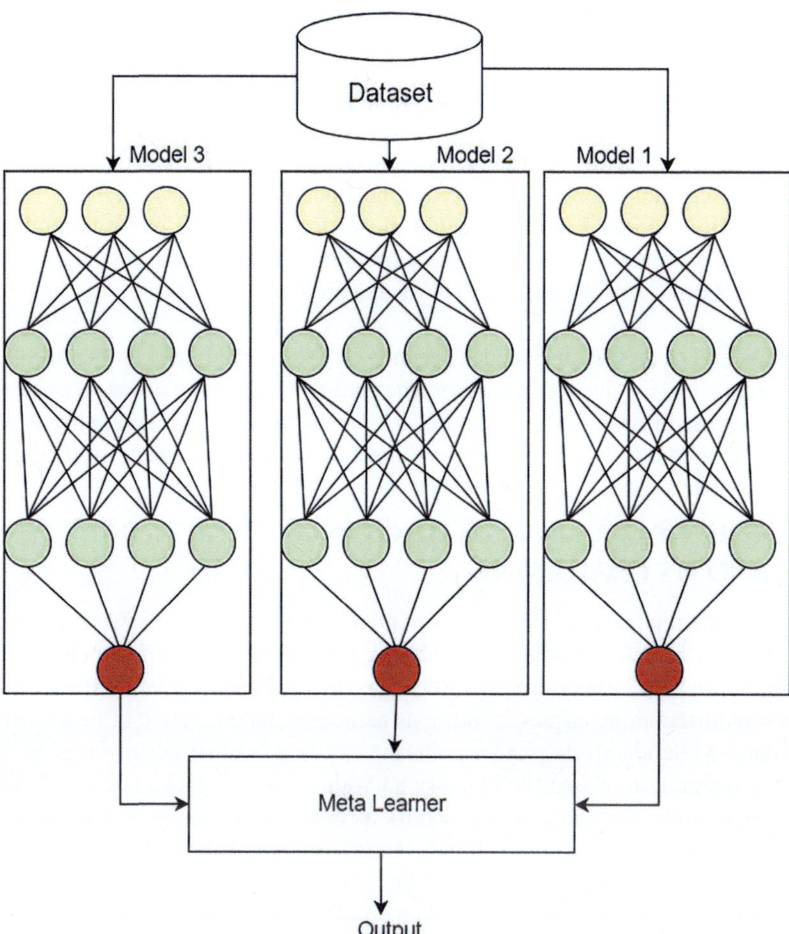

FIGURE 2.6 Ensemble model architecture.

- **Stacking:** Stacking, a distinctive method, melds the predictions of multiple models through the utilization of a meta-model. The meta-model acquires the proficiency to amalgamate the predictions of individual models in a manner that accentuates their collective precision.

j. **Hybrid Models:** Hybrid models represent a class of machine learning frameworks that harness the advantages of diverse model types. These models harmonize the capabilities of deep learning methodologies with conventional climate models, thereby capitalizing on the merits of both methodologies. The outcome is the potential for heightened precision and comprehensibility in predictions and risk evaluations. This synergy emerges as a potent instrument for climate change prediction, as it offers the potential to counterbalance the limitations of standalone models. Various approaches exist for integrating disparate model types within a hybrid framework [14]. Among the most frequently employed techniques are:

- **Transfer Learning:** Transfer learning trains a model on a related problem and then uses that model as a starting point for training a model on the target problem. This can facilitate an enhancement in prediction accuracy by capitalizing on the knowledge the model has already acquired.
- **Multi-Model Learning:** Multi-model learning trains multiple models on different aspects of the same problem. This can help to improve the robustness of the models to noise and outliers.

The choice of deep learning model depends on the specific task, the nature of the climate data, and the desired outcomes. The quality and quantity of data are essential for the success of deep learning models. Climate scientists and data analysts often experiment with different models and architectures to find the best fit for their goals.

2.4 ROLE OF DATA IN CLIMATE CHANGE PREDICTION AND ITS CHALLENGES

Deep learning models necessitate a substantial volume of data for effective training. The selection of data suitable for leveraging deep learning in climate change prediction and risk assessment hinges on the specific application at hand. For instance, projecting forthcoming climate conditions mandates the utilization of historical climate data, while identifying and monitoring extreme weather occurrences necessitates the integration of satellite imagery and other relevant data sources. Modeling the repercussions of climate change across diverse sectors demands a multifaceted dataset, encompassing historical climate records, societal data, and sensor-generated data. The efficacy of deep learning in climate change prediction and risk assessment hinges on the amalgamation of varied data types, enabling the encapsulation of intricate interactions within the Earth's climate system and the evaluation of potential ramifications. Several categories of data that can be harnessed include:

a. **Meteorological and Climate Data:**
   This data serves as training material for instructing deep learning models to discern the interconnections among diverse climate parameters. Subsequently, these models can be harnessed to prognosticate forthcoming climate circumstances, encompassing elements like temperature, precipitation, and sea level escalation.

   - **Temperature:** Surface and air temperature records from the past are essential for studying climate change.
   - **Precipitation:** Precipitation records are useful for monitoring climate phenomena like droughts and floods.
   - **Wind Data:** Knowing how fast and which way the wind is blowing is crucial for analyzing weather systems and predicting their movements in the atmosphere.

- **Humidity:** Humidity data are used to study atmospheric moisture distribution, climate, and surroundings.

b. **Satellite Imagery Data:** Satellite images are a great resource to monitor changes in land usage, ice extent, and sea level. This data is crucial for understanding climate change and predicting future changes.

- **Visible Imagery Data:** Visible satellite photography is the most basic one in which the human-visible light spectrum is recorded. Visible pictures can identify forests, grazing land, and cities. Both deforestation and desertification can be used as indicators of flora change.
- **Infrared Imagery Data:** Infrared photography records the range of the electromagnetic spectrum beyond the range of the human eye. Infrared imaging can detect temperature changes brought on by heat waves, droughts, and wildfires. It can also detect changes in plant life, such as stress levels and water status, that would otherwise go unnoticed.
- **Data from Multispectral Imaging:** Images captured by a multispectral camera capture both visible and infrared light at different wavelengths. Changes in land use and vegetation may now be studied in finer detail. In order to keep track of things like crop growth and water stress in agricultural settings, multispectral photography is frequently employed.

c. **Biodiversity and Ecosystem Data:** Knowledge of ecosystems and biodiversity is essential for identifying vulnerable areas and developing appropriate adaptation strategies in the face of climate change. Considerable use may be made of the following ecological and biodiversity information in the context of climate change:

- **Distribution of Species:** Researching how species' ranges shift throughout time requires information on where certain species now live. The effects of climate change on biodiversity may be tracked and important habitats for endangered species can be identified with the use of this data.
- **Ecosystem Functioning:** Carbon sequestration and nitrogen cycling are only two examples of how changes in ecosystem functions may be tracked with the use of data on ecosystem functions. Insights into how these findings affect ecosystems' capacity to provide human needs in the face of climate change are now possible.

d. **Ocean Data:** Due to the ocean's central role in controlling global temperature, accurate predictions of climate change's consequences rely heavily on ocean data. The seas absorb around 30 percent of the carbon dioxide produced by humans, which helps to keep the planet at a constant temperature [15]. Oceanographic information often takes the following forms:

- **Ocean Salinity:** Information on the effects of salinity fluctuations on ocean circulation and the distribution of marine life might be gleaned through monitoring these fluctuations. The presented information may be used to foretell how the ocean's ability to moderate global temperatures would develop in the future.

- **Sea Surface Temperature (SST):** Data on SST may help scientists track how other climate variables, including precipitation and evaporation, react to changes in sea levels. Using this information, forecasters can anticipate weather patterns and extreme weather.
- **Ocean Current:** Scientific understanding of the ocean's heat and nutrient movement may be improved using data collected through observations of ocean currents. Using this information, we may speculate on how the ocean's ability to affect global climate may change in the future.
- **Sea Level Rise Data:** Data on sea level rise is useful for tracking how the rising seas can affect coastal communities and ecosystems. Communities located near the coast may use this information to prepare for the future impacts of climate change.

e. **Atmospheric Composition Data:** This can shed light on how shifts in atmospheric composition are influencing the planet's weather. Greenhouse gases, ozone, and other air pollutants can all be tracked using this information. The following are some instances where atmospheric composition data has been utilized to detect climate change:

- **Greenhouse Gas Concentration Data:** Greenhouse gases are a major contributor to global warming because they trap heat in the atmosphere. Information on atmospheric composition is essential for tracking changes in greenhouse gas concentrations and predicting future climatic changes on Earth.
- **Ozone Levels Data:** Ozone is a gas that absorbs harmful ultraviolet radiation from the sun. Data pertaining to atmospheric composition, specifically ozone levels, is instrumental in monitoring variations in these levels and anticipating the prospective repercussions of climate change on both human well-being and the environment.
- **Air Pollutant Concentrations:** Airborne pollutants can significantly contribute to respiratory issues, heart ailments, and various health complications. Atmospheric composition data concerning concentrations of air pollutants offers a valuable resource for monitoring shifts in these pollutants and projecting the forthcoming ramifications of climate change on human health.

f. **Social and Economic Data:** Social and economic data is also important for climate change prediction because it can help us to understand how human activities are affecting the Earth's climate. This data can be used to identify trends in population growth, energy use, and economic development.

- **Population Growth:** Population expansion places immense pressure on resources and culminates in escalated emissions of greenhouse gases. Social and economic data on population growth can be used to predict future emissions and to develop strategies for mitigating climate change.
- **Energy Use:** Energy consumption serves as a significant origin of greenhouse gas emissions. Social and economic data pertaining to energy usage plays a pivotal role in prognosticating forthcoming emissions and in devising strategies to curtail energy consumption.

- **Economic Development:** Economic development can lead to increased emissions of greenhouse gases. Social and economic data on economic development can be used to predict future emissions and to develop strategies for decoupling economic growth from emissions growth.

Integrating and analyzing these diverse data sources using deep learning techniques can provide a comprehensive understanding of climate dynamics, predict future trends, and assess potential risks and vulnerabilities associated with climate change. But there are several challenges in obtaining and preparing data for deep learning for climate change prediction and risk assessment. These challenges include:

a. **Data Availability:** The availability of data is a key challenge for deep learning for climate change prediction and risk assessment. Some data types, such as historical climate data, are relatively easy to obtain. However, other data types, such as satellite imagery and social data, can be more difficult. In addition, the quality of data can vary greatly. Ensuring the accuracy and reliability of deep learning models necessitates meticulous curation of the data employed for training.

b. **Data Quality:** The quality of data can also be a challenge for deep learning for climate change prediction and risk assessment. Some data types, such as satellite imagery, can be noisy and contain errors. Before utilizing data for training deep learning models, it holds paramount importance to undertake data cleaning and pre-processing, aiming to eliminate noise and rectify errors.

c. **Data Imbalance:** Data imbalance is another challenge for deep learning for climate change prediction and risk assessment. This occurs whenever there is a large discrepancy in the sample size between categories. To illustrate, there might be a substantially greater number of data samples representing normal weather conditions compared to the relatively scarce samples depicting extreme weather events. This can lead to deep learning models that are biased towards predicting normal weather conditions. Before using data to train deep learning models, it is important to address data imbalances by either over- or under-sampling the data.

d. **Data Labelling:** Data labelling is the process of assigning labels to data points. This can be a challenging task for climate change data, as it can be difficult to determine the labels for some data points. For example, it can be difficult to determine whether a particular weather event is an extreme weather event or not. It is critical to carefully label the data to ensure that the deep learning models are trained on accurate data. Mislabeling or inaccurate ground truth can adversely affect model performance.

e. **Spatial and Temporal Resolution:** Integrating data with different spatial and temporal resolutions can be challenging and may require interpolation, aggregation, or downscaling methods. Balancing the trade-off between data resolution and computational complexity is important.

f. Data Integration and Fusion: Merging data originating from diverse sources, encompassing satellite imagery, climate models, and ground observations, demands the utilization of data integration methodologies. The intricate process involves ensuring harmonious integration and preserving data integrity throughout the procedure.

e. Data Cost: The cost of obtaining and preparing data can also be a challenge for deep learning for climate change prediction and risk assessment. Some data types, such as satellite imagery, can be expensive to obtain. In addition, the cost of cleaning and pre-processing data can also be significant. It is important to consider the cost of data when developing deep learning models for climate change prediction and risk assessment.

Notwithstanding these hurdles, deep learning remains a propitious instrument for the realms of climate change prediction and risk assessment. With the ongoing advancement of deep learning technology and the enhancement of data accessibility, we can anticipate witnessing further ingenious applications of this technology in addressing the complexities of climate change.

2.5　APPLICATION OF DEEP LEARNING IN CLIMATE CHANGE PREDICTION AND RISK ASSESSMENT

Deep learning techniques can be utilized for the purpose of climate change prediction and risk assessment. These approaches involve the analysis of massive and complex datasets to identify patterns, trends, and insights linked to the dynamics of the climate and prospective repercussions. Some ways in which deep learning can be applied in this context are:

a. Climate Modeling and Prediction: The use of mathematical models to mimic the climate system is known as "climate modeling," which is also used in climate prediction. Models of the climate are crucial for making accurate forecasts and gaining insight into past and present variations. This understanding is instrumental in formulating efficacious strategies for both adaptation and mitigation. Climate models constitute an invaluable asset in comprehending and prognosticating climate change, offering insights into scenarios such as:

- **Weather and Climate Forecasting:** Deep learning can enhance the accuracy of weather and climate predictions by analyzing historical climate data and real-time observations. RNNs and LSTM networks can capture temporal dependencies in climate data, leading to improved short-term and long-term forecasts.
- **Extreme Weather Event Prediction:** Deep learning models can undergo training to recognize intricate patterns linked to extreme weather phenomena, including hurricanes, heatwaves, and floods. This capability equips them to deliver predictions characterized by heightened accuracy and timeliness, thereby contributing to improved readiness and response measures.

b. **Climate Data Analysis and Pattern Recognition:** Climate data analysis encompasses the collection, arrangement, and interpretation of climate-related data. This information is instrumental in monitoring climate shifts over temporal spans, discerning trends, and formulating future projections. On the other hand, pattern recognition involves the identification of patterns inherent in data. This process facilitates the recognition of interconnections between diverse climate variables, like temperature, precipitation, and sea level. Essential techniques in climate data analysis encompass feature extraction and climate data fusion.

- **Feature Extraction:** Deep learning algorithms possess the capacity to autonomously extract pertinent features from expansive and multi-dimensional climate datasets, thereby unveiling concealed patterns and correlations. Principal Component Analysis (PCA) has proven effective in distilling the most pivotal features from climate datasets encompassing parameters like temperature, precipitation, and sea level. This application has contributed to refining the precision of climate models and unearthing intricate patterns within climate data.

- **Climate Data Fusion:** Climate data fusion involves the amalgamation of distinct climate datasets into a coherent and singular dataset. This harmonization is accomplished through diverse methodologies, encompassing data assimilation, ensemble learning, and machine learning techniques. Data assimilation has been effectively employed to merge disparate climate datasets, ranging from satellite- and ground-based observations to model-generated data. This addition has substantially enhanced the realism of climate models and offered a more comprehensive view of the climate system. Deep learning, in particular, has the ability to unify different types of climate data, such as satellite photos, sensor data, and computer models of the climate, into a cohesive whole, resulting in a more thorough and accurate knowledge of climate dynamics.

c. **Risk Assessment and Impact Analysis:** The impact analysis looks at how people, ecosystems, and buildings are affected by climate change, while the risk assessment looks at the possibility and consequences of different climate change scenarios. The typical methods used to evaluate the dangers posed by climate change are as follows:

- **Identifying Climate Change Hazards:** Risks associated with climate change, such as droughts, floods, heat waves, and storms, must be identified before any action can be taken to mitigate them.

- **Estimating the Likelihood of Hazards:** Predicting the possibility of specific threats occurring under different climatic scenarios requires first using climate models to forecast probable future climate circumstances, and then use statistical methodologies to make such predictions.

- **Assessing the Consequences of Hazards:** Possible effects, such as the number of people who would need to be evacuated due to a flood or the amount of damage a storm may bring to buildings and other infrastructure, can be evaluated via hazard assessment.

Climate change impact analysis typically involves the following steps:

- **Identification of Climate Change Implications:** The first step in adapting to a changing climate is identifying the consequences of that shift for people, ecosystems, and man-made structures.
- **Impact Extent Estimation:** To estimate the full scope of an influence, researchers utilize impact models to examine the potential consequences of a range of climate change scenarios.
- **Demographic Vulnerabilities to Climate Change:** Particular people or ecosystems may be more susceptible to the consequences of climate change; thus, it is essential to identify those at risk.
- **Strategizing Adaptation:** Adaptation strategy involves developing plans to protect vulnerable populations from climate change's negative effects.

Examples of where deep learning has been used to evaluate climate change risk and effect include:

- **Sea Level Rise and Coastal Flooding:** Increased flooding along coastlines due to rising seas may be predicted with the use of deep learning algorithms that analyze satellite photos and previous data. With this information, infrastructure improvement and expansion plans may be made.
- **Crop Yield Prediction:** Using temperature data, soil conditions, and past crop yield data, deep learning can predict agricultural outcomes and evaluate the possible effect of climate change on food production.
- **Impact on Ecosystems and Biodiversity:** Deep learning has the potential to aid in assessing the consequences of climatic changes on ecosystems and biodiversity by analyzing satellite photos, data on the distribution of species, and climate models.

d. Climate Change Attribution: Climate change attribution involves the process of determining the degree to which specific climate change events or trends are a result of human activities. Utilizing deep learning, one can engage in climate change attribution by recognizing the climate data patterns that are likely attributed to human actions. This distinction aids in distinguishing between natural climate fluctuations and alterations caused by human influence. Deep learning offers several avenues for climate change attribution. One approach entails training a deep learning model to forecast climate outcomes devoid of human impact. This model can subsequently be employed to juxtapose observed climate data with projected data, pinpointing patterns potentially caused by human activities. An alternative method involves leveraging deep learning to unveil the causal connections among diverse climate variables. Accomplished by training a model to predict one climate variable based on another, this process helps identify underlying causal relationships between the two variables.

e. Early Warning Systems: Early alert systems hold a crucial role in adapting to climate change, facilitating readiness and reactions to severe weather incidents, encompassing floods, droughts, and heatwaves. Deep learning, a formidable instrument, offers the potential to enhance the precision and promptness of climate change early warning systems. Various avenues exist through which deep learning can bolster early warning systems for climate change:

- **Detecting Early Indicators:** Deep learning algorithms can be trained on historical climate records to discern patterns suggestive of an upcoming extreme weather occurrence. For instance, a deep learning model could learn to recognize alterations in precipitation frequency or intensity, which could signal an impending flood.
- **Anticipating Timing and Location of Severe Weather Events:** Deep learning models are also applicable for forecasting when and where extreme weather events might transpire by leveraging past data. This predictive insight can facilitate the issuance of timely alerts to regions at risk.
- **Conveying Advance Notices to Affected Communities:** Deep learning models have the capacity to generate customized early warning messages tailored to the specific requirements of impacted communities. This tailored approach ensures that crucial early warnings are received and acted upon by those who stand to benefit the most.

The proficiency of deep learning in handling extensive and intricate datasets, extracting intricate patterns, and producing precise forecasts establishes it as a valuable asset in the realms of climate change investigation, risk evaluation, and mitigation endeavors. Nonetheless, it remains essential to acknowledge that the utilization of deep learning methodologies should be supplemented by domain expertise, conventional climate models, and pertinent data. This comprehensive approach guarantees the attainment of resilient and dependable outcomes.

2.6 CONCLUSION

Deep learning has emerged as a potent instrument for prognosticating climate change and evaluating associated hazards. Its intricate models can comprehend intricate relationships among climate parameters, enabling projections of forthcoming climate conditions. These projections hold the potential to guide policy formulation and assist communities in readying themselves for climate change's repercussions.

Nonetheless, several challenges must be confronted to effectively employ deep learning for climate change prediction and risk assessment. One obstacle lies in the substantial data requirements for training deep learning models. Accumulating this data, particularly for long-term climate forecasts, can prove arduous and costly. Moreover, the computational demands of training and operating deep learning models present another hurdle, potentially limiting their applicability in developing nations and remote locales. Despite these impediments, the promise of deep learning in enhancing climate predictions and identifying vulnerable regions remains substantial. This knowledge can underpin policy decisions and support community preparedness in the face of climate change impacts.

REFERENCES

1. J. T. Shaw, G. Allen, D. Topping, S. K. Grange, P. Barker, J. Pitt, and R. S. Ward. "A case study application of machine-learning for the detection of greenhouse gas emission sources," Atmospheric Pollution Research, vol. 13, no. 10, 2022, 101563.

2. "Paris Agreement to the United Nations Framework Convention on Climate Change, Dec. 12, 2015, T.I.A.S. No. 16–1104." Accessed on: Aug. 28, 2023. [Online]. Available: https://guides.brooklaw.edu/c.php?g=1197665&p=8776472

3. E. Plane, K. Hill, and C. May, "A Rapid assessment method to identify potential groundwater flooding hotspots as sea levels rise in coastal cities," *Water (Switzerland)*, vol. 11, no. 11, Nov. 2019, doi: 10.3390/w11112228.

4. "NOAA scientists harness machine learning to advance climate models." Accessed on: Aug. 28, 2023. [Online]. Available: *www.gfdl.noaa.gov/news/noaa-scientists-harness-machine-learning-to-advance-climate-models/*

5. L. Alzubaidi *et al.*, "Review of deep learning: concepts, CNN architectures, challenges, applications, future directions," *J Big Data*, vol. 8, no. 1, Dec. 2021, doi: 10.1186/s40537-021-00444-8.

6. J. Kim, M. Lee, H. Han, D. Kim, Y. Bae, and H. S. Kim, "Case study: development of the CNN model considering teleconnection for spatial downscaling of precipitation in a climate change scenario," *Sustainability (Switzerland)*, vol. 14, no. 8, Apr. 2022, doi: 10.3390/su14084719.

7. J. Kang, H. Wang, F. Yuan, Z. Wang, J. Huang, and T. Qiu, "Prediction of precipitation based on recurrent neural networks in jingdezhen, jiangxi province, China," *Atmosphere (Basel)*, vol. 11, no. 3, Mar. 2020, doi: 10.3390/atmos11030246.

8. D. M. Jose, A. M. Vincent, and G. S. Dwarakish, "Improving multiple model ensemble predictions of daily precipitation and temperature through machine learning techniques," *Sci Rep*, vol. 12, no. 1, Dec. 2022, doi: 10.1038/s41598-022-08786-w.

9. P. Guan, L. Zhu, and Y. Zheng, "A study of forest phenology prediction based on GRU models," *Applied Sciences (Switzerland)*, vol. 13, no. 8, Apr. 2023, doi: 10.3390/app13084898.

10. G. Czibula, A. Mihai, A. I. Albu, I. G. Czibula, S. Burcea, and A. Mezghani, "Autonowp: An approach using deep autoencoders for precipitation nowcasting based on weather radar reflectivity prediction," *Mathematics*, vol. 9, no. 14, Jul. 2021, doi: 10.3390/math9141653.

11. J. Cheng *et al.*, "DeepDT: Generative adversarial network for high-resolution climate prediction," *IEEE Geoscience and Remote Sensing Letters*, vol. 19, 2022, doi: 10.1109/LGRS.2020.3041760.

12. J. Xu, H. Fan, M. Luo, P. Li, T. Jeong, and L. Xu, "Transformer based water level prediction in Poyang Lake, China," *Water (Switzerland)*, vol. 15, no. 3, Feb. 2023, doi: 10.3390/w15030576.

13. L. Hernan, L. Martí, N. Sanchez-Pi, and H. Lira, "Frost forecasting model using graph neural networks with spatio-temporal attention." [Online]. Available: https://hal.inria.fr/hal-03259658

14. L. J. Slater *et al.*, "Hybrid forecasting: blending climate predictions with AI models," *Hydrology and Earth System Sciences*, vol. 27, no. 9, May 15, 2023, 1865–1889, doi: 10.5194/hess-27-1865-2023.

15. "Effects of changing the carbon cycle." Accessed on: Aug. 28, 2023. [Online]. Available: https://earthobservatory.nasa.gov/features/CarbonCycle/page5.php

3 AI- and IoT-Based Applications for Rainfall Prediction
A Study

*Manoj Kumar Pandey, Sunil Kumar Singh,
Jyoti Upadhyay, Preeti Tiwari,
Naresh Kumar Kar, and Jai Prakash Kushwaha*

3.1 INTRODUCTION

Rainfall prediction is one of the most important aspects for everyone, and usages of artificial intelligence (AI) and Internet of Things (IoT) may impact the overall process of prediction. Rainfall prediction is very important for several reasons, as making good predictions impacts the agriculture to a great extent. The scarcity of rainfall has a negative impact on the aquatic ecosystem, water supply, and productivity. The rainfall prediction is based on various factors, such as wind speed, moisture, humidity, temperature, pressure, min-temp etc., and these parameters can be very useful to predict the rainfall by applying various machine learning models. These predictions are based on datasets belonging to regions and bounds to time duration and rainfall is predicted by applying various machine learning models. The time duration of data collection also affects the overall performance of rainfall prediction; assuming one can get real-time data for training purposes, it will improve the model performance and prediction can be more effective and accurate. IoT enables all of us to collect real-time data using various sensors designed for specialized tasks. These sensors are installed and are well connected with the cloud servers so that real time data can be collected easily. This chapter discusses various aspects of rainfall prediction using AI and IoT.

A transformative era has recently begun due to the convergence of Artificial Intelligence (AI) and the Internet of Things (IoT), which has completely changed how data is gathered, examined, and applied to solve complicated problems. Predicting rainfall is a crucial field with broad consequences for disaster planning, agricultural production, water resource management, and other issues. This synergy has several intriguing applications in this field. This dynamic combination of AI and IoT technologies has made it possible to create complex systems that can provide accurate forecasts in a timely manner, improving our capacity to

DOI: 10.1201/9781003452393-3

foresee and react to changes in precipitation patterns [1]. The development of artificial intelligence has significantly altered how we examine enormous and complex datasets, revealing patterns and links that were previously hidden. AI approaches enable the extraction of useful insights from a wide range of meteorological variables, historical data, and real-time observations collected by IoT devices in the context of rainfall prediction. Few machine learning algorithms such as decision trees [2], K-nearest neighbour [3], support vector machines, and neural networks [4] are examples of machine learning algorithms that independently learn from previous data and continuously improve their predictions through iterative learning processes. As a result, a prediction framework that can recognize the complex, nonlinear interactions that frequently control rainfall patterns is created. Arabelli et al., 2023, [5] has employed decision trees (DT), logistic regression (LR), support vector machine (SVM), and random forests (RF) to perform the weather prediction; for collecting real-time data, various sensors like DHT11 and BMP180 are used, and one module, ESP8266, is used to add the sensor generated data to the database. Here, the random forest models show the highest accuracy of 84% and take 2.4 seconds.

It is obvious that recent advancements in the IoT space have given us all the power to gather real-time information on the forecasts for rainfall, and using that information to make a prediction using a machine learning approach can provide some truly amazing outcomes. The datasets, pre-processing method, and models chosen for rainfall prediction all play a major role in forecasts. To effectively plan for the country and to ensure that standards of life are maintained, it is crucial to anticipate rainfall with great accuracy. Predicting rainfall has many benefits for nations, benefiting different societal sectors and facets. Its contribution to agricultural planning and food security is a notable advantage. Accurate rainfall forecasts enable farmers to make informed decisions about planting, irrigation, and harvesting, optimizing crop yields, and ensuring a stable food supply for the population [6].

Additionally, rainfall prediction plays a pivotal role in water resource management. By anticipating precipitation patterns, authorities can effectively manage reservoir levels, groundwater recharge, and water distribution for various uses such as domestic consumption, agriculture, and industrial processes. This contributes to sustainable water management and helps mitigate the impact of water scarcity [7]. By offering early warning systems for catastrophic weather occurrences like floods and landslides, the projections also improve disaster preparedness. Initiating evacuation preparations, allocating resources, and coordinating emergency response activities are all made possible by prompt alerts based on rainfall projections, which lowers the loss of life and damage.

Accurate rainfall forecasting is also important for infrastructure planning. With the help of projected rainfall patterns, urban planners and engineers may create robust drainage systems and flood management plans. By ensuring infrastructure longevity and lowering the risk of urban floods, these measures eventually lead to lower maintenance costs [8]. The ability to predict rainfall is crucial for energy generation, particularly in hydroelectric power plants. Power generation is directly impacted by the availability of water in reservoirs. Energy companies can optimize their operations, control reservoir levels, and guarantee a reliable and efficient energy supply with the help of precise forecasts [9].

Forecasting rainfall benefits economic stability across a range of industries. Businesses in industry, tourism, and agriculture can modify their plans based on the forecasted weather, reducing losses and maximizing income [10]. Prediction of rainfall has health effects as well. Authorities can take action to prevent water stagnation and the creation of breeding grounds for disease-carrying insects like mosquitoes by planning beforehand for heavy rain. By doing so, public health is protected and vector-borne diseases are controlled. Insights gained from precise rainfall forecasting also aid in the study and modelling of the climate [11].

3.2 RELATED WORK

This section deals with the recent studies done for predicting rainfall using IoT devices and machine learning methods. The purpose of this section is to familiarize the reader about current trends in this research. After looking at various reviews of the literature, it is observed that there are lots of scholarly articles related to IoT and rainfall, but only articles from the last six to seven years have been considered so that this study deals with some of the more recent significant work.

Shah et al. 2023 [12] have done a study to predict the rainfall of three regions i.e., Chennai, Calicut, Vadodara. To collect the latest rainfall data, IoT devices have been employed and kept in these three cities. IoT devices DHT 11 and BMP 180 have been controlled using microcontroller ESP8266 and interfaced with the PowerBi, so that collected data can be easily managed and shared. Regression and classification algorithms are used to predict rainfall in the city. Decision tree regression and random forest regression algorithms have been used for the regression analysis and decision tree classifier and Naïve Bayes classifiers have been used for classification. Along with the dataset collected using the IoT devices, another standard database has been used, and the performance of various machine learning models using these two databases has been compared. It is found that a decision tree regression model gives an accuracy of 84.73% for raw data and accuracy of 88.23% for IoT-based datasets. Random forest regression shows accuracy of 85.20% for raw data and accuracy of 85.68% for IoT-based datasets. For the classification task, the decision tree classifier achieves an accuracy of 92% for raw data and accuracy of 95% for IoT-based datasets. In the case of the Naïve Bayes classifier, it achieves an accuracy of 81% for raw data and accuracy of 83% for the IoT-based dataset. These results and the conclusion of the study show that there is a difference of 2–3% accuracy, i.e., the IoT-based dataset gives an improved accuracy of 2–3% compared to raw dataset-based accuracy. It also shows that using IoT for recording the data related to rainfall may lead to better results.

Liyew et al. 2021 [13] conducted yet another investigation to pinpoint the pertinent atmospheric elements that cause rain to fall and to forecast the daily severity of rainfall using machine learning techniques. Three different machine learning models, including multivariate linear regression (MLR), random forest (RF), and extreme gradient boost (XGBoost), were used to collect data from Ethiopia's local meteorological agency. For the purposes of the study, a 20-year-old raw dataset containing 10 data features—including year, month, date, evaporation, daylight, maximum temperature, minimum temperature, humidity, wind speed, and rainfall—was taken into consideration. Mean value replacement has been used to deal with the missing data. The full datasets have been partitioned into 80% training and 20% testing. Pearson correlation is also applied to find the significant features for rain prediction, and it is found that humidity and wind speed are the significant features for rainfall prediction, with a Pearson coefficient value of 0.402 and 0.351 (greater than 0.2). The rainfall prediction is performed using all three mentioned machine learning algorithms, and it is found that XGBoost predicted the rainfall better than the RF and MLR models. The author concluded that this result may be further improved if sensor data is incorporated for the study.

In order to make accurate predictions, Sathya et al. 2023 [14] suggested an Enhanced Learning Scheme for Weather Prediction (ELSWP), which is based on the traditional machine learning logistic regression (LR) model and used IoT device DHT-11 and a rainfall sensor. However, it has a drawback in that it can occasionally function poorly; the data collected by IoT devices only covers a small area.

Sharma et al. (2022) [15] have presented a study for the forecast of rainfall for farmers, and for that purpose a 'SMART CAP' was built, which is wearable and comprises microprocessors that can capture atmosphere parameters with the assistance of several sensors. It can be worn by farmers as they labour so that they can record the temperature and humidity. With the aid of a WiFi module, the recorded value is then entered into "THINGSPEAK" online. The prediction accuracy using the linear and logistic regression models is 81% and 80.23%, respectively.

Maliyeckel et al. (2021) [16] have performed a study to measure the performance of machine learning algorithms like light gradient-boosting machine (LightGBM) and support Vector Regression (SVR), and they have also developed an ensemble model based on LightGBM and SVR to predict rainfall. In order to accomplish this, they have utilised the database with three features: humidity, wind-speed, and temperature. A rainfall database from the period of 1st January 2019 to 31st August 2020 is considered for the training, and the database from the period of 1st September 2020 to 30 September 2020 is considered the testing dataset. Missing data is managed using the mean value. The Pearson correlation coefficient is used to find the linear relation between the two variables, and after evaluation it was found that all the parameters are significantly important to rainfall. The Mean Absolute Error (MAE) evaluated for the LightGBM is 5.73, and for SVR MAE is 8.78. However, the hybrid model shows an MAE of 4.90, which is more efficient than the other classical model.

Another study, by Dash et al. (2018) [17], used three machine learning techniques—Artificial Neural Network (ANN), K-nearest neighbour (KNN), and Extreme Learning Machine (ELM)—to estimate rainfall in the Kerala state from 2011 to 2016. The last 47 years of data, separated into the months of January–February, March–May, June–September, and October–December, were used to train the machine learning models. Every piece of training data contains several statistical metrics, such as mean, median, mode, lowest, and maximum. System performance is gauged using Root Mean Square Error (RMSE), Mean Absolute Error (MAE), Standard Deviation (SD), and Mean. The addition of the hidden layer has a greater impact on the outcome for ANN and ELM, and the 8–15–1 ELM architecture exhibits positive outcomes. The experimental finding indicates that, with an MAE of 3.075% for summer monsoon datasets and an MAE of 3.149% for post-monsoon datasets, the ELM technique has outperformed the other two strategies.

Another study of rainfall prediction is done by Huang et al. (2017) [18], which was totally based on the K-nearest neighbour algorithm. In this work, improved KNN is utilized for rain prediction over three other methods: namely, distance weighted K-nearest neighbour (WKNN) and dual weighted K-nearest neighbour (DWKNN), KNN. Here the training sample used is 1748 records, and testing samples contain 368 records. The proposed improved KNN shows the better performance score of 50.46% for K = 6 against the three other algorithms, namely DWNN, WKNN, and KNN with scores of 48.32%, 48.62%, and 46.88%.

In a study conducted by Chao et al. in 2018 [19], rainfall was predicted utilizing a micro-electromechanical systems (MEMS) sensor to store relevant data for waterfall prediction and a long short-term memory to forecast real-time data based on data collected. Numerous machine learning-based techniques, including support vector machine (SVM), autoregressive and moving average (ARMA), random forest (RF), and back propagation neural network, are compared to the predicted data. MEMS sensors [20] do have complex mistakes; however, the manufacturer may remove them before the sensors are sold by employing calibration methods. In order to monitor the pertinent data, four Wuhan stations have been located, and seven sensors have been installed. The seven sensors are a rainfall sensor, a humidity sensor, a pressure sensor, a pressure sensor for wind, a temperature sensor, and a radiation sensor. The LSTM is seen to perform better than other machine learning models.

3.3 COMPARISON OF PREVIOUS WORK DONE

In this section, some recent work as shown in Table 3.1 has used various parameters like literature, ML method used, IoT device used, Pros and Cons, and Accuracy/MAE in comparison to the previous work done on rain prediction using IoT devices and machine learning methods.

Arabelli et al. 2023 [5] have applied various machine learning methods like logistic regression, SVM, decision tree, and random forest for predicting the rain, and data has been collected using various IoT-based devices that achieved accuracy of 83.9%, 83.6%, 76%, and 84.2% for various machine learning-based methods. Google Collab is used in the proposed approach.

Shah et al. 2023 [12] have predicted rain prediction using various machine learning methods such as decision tree and random forest using Naive Bayes classifiers; real-time data has been collected using IoT devices, achieving accuracy of 95% for decision tree and 83% for Naive Bayes classifiers. In this work, PowerBi is used for the data visualizations.

Another rain prediction was done by Sharma et al. 2022 [15], in which multiple linear regression is used for rain prediction, and wearable devices are used for data collection. This achieved accuracy of 81% for the multiple linear regression method. This method is expensive, as wearable devices are expensive than ordinary IoT devices.

Another rain prediction has been done by Xu et al. 2022 [16], in which a standard dataset was utilized for rain prediction. In this, machine learning methods like LSTM were utilized, and PSO was used for data optimization, achieving accuracy of 94%. Another method for rain prediction has been proposed by Maliyeckel et al. 2021 [17], and standard datasets were selected; machine learning methods like LightGBM, SVR, and Hybrid models were utilized, and Mean Square Error obtained was 5.73, 8.78, and 4.90 respectively for machine learning methods.

Another method of rain prediction has been proposed by Emmanuel et al. 2021 [21]. The standard dataset for rain prediction was used along with various machine learning methods like Artificial Neural Networks (ANN)—Feed Forward Neural Network (FFNN), Cascade Forward Neural Network (CFNN), Recurrent Neural Network (RNN), and Elman Neural Network (ENN), but this method takes more time and is more expensive, as deep learning was applied to this. The lowest Mean Square Error ENN gives a good performance for 2018.

Another method for rain prediction has been proposed by Shalini et al. 2021 [22], using a geometry algorithm on an IoT platform. It causes a time delay but provides an accuracy rate of 100%. Another method for rainfall prediction using machine learning has been proposed by Zhao et al. 2021 [23], in which various machine learning models, such as linear regression (LR), Support Vector Classification, K-nearest neighbour, Gaussian Naïve Bayes, Bernoulli Naive Bayes, AdaBoost, Gradient Boosting, Bagging, random forest, Extra Trees, Gaussian Process Classification, logistic regression, Extra Tree, decision tree, and Quadratic Discriminate Analysis, were used. Here, no IoT device is utilized; instead, water flow of debris is used for creating the datasets, and for Quadratic Discriminate Analysis it provides the highest accuracy of 100%.

Another method for rain prediction has been proposed by Sadhukhan et al. 2021 [24], in which various machine learning methods, such as SVM, KNN, DNN, and ANN, have been applied and various IoT devices are utilized for reading the data related to rainfall, providing an accuracy rate of 79.6%, 70%, 84%, and 89%, respectively. The advantage of this system is that it is accurate, portable, and reliable.

Another method for rainfall prediction has been proposed by Rani et al. 2020 [25], in which various IoT devices are used with machine learning methods like linear regression, logistic regression, support vector machine, and ANN; for SVM the achieved accuracy was 90.60%, and for ANN, 88%.

Another rainfall prediction using an SVM model is proposed by Onkar et al. 2019 [26]. This method is flexible and efficient but requires a time delay. Another method for rainfall prediction has been proposed by Mzyece et al. 2018 [27], in which ANN is utilized with a standard image dataset and it achieved accuracy of 99%.

TABLE 3.1

Comparison of Previous Work Done

Work done	ML method used	IoT device used	Pros and Cons	Accuracy/MAE
Arabelli et al. 2023 [5]	Logistic regression, SVM, decision tree, random forest	DHT-11, BMP-180	Real time data, Execution time Google Colab is used	LR – 83.9% DT – 76% RF – 84.2% SVM – 83.6%
Shah et al. 2023 [12]	Decision tree regression, random forest regression, decision tree classifier, naïve Bayes classifiers	DHT-11, BMP-180	Real time data Efficient Use of power Bi	95% for Decision tree 83% for Naïve Bayes
Sathya, et al. 2023 [14]	Logistic regression (LR)	DHT11	Effective Real time data Work on Limited data	LR – 92%
Sharma, et al. 2022 [15]	Multiple linear regression, logistic regression	Microcontroller (Arduino UNO), LM 35 Sensors	Device is wearable Expensive	MLR – 81%
Xu et al., 2022 [16]	LSTM, PSO for optimization	NA	Optimized Accuracy Time taking	LSTM – 94%
Maliyeckel et al. 2021 [17]	LightGBM, SVR, Hybrid model	None	Efficient Hybrid model is applied Large scale data is missing	For LightGBM MAE is 5.73 For SVR MAE is 8.78 For Hybrid model MAE is 4.90
Emmanuel et al. 2021 [21]	Artificial Neural Networks (ANN) – Feed Forward Neural Network (FFNN), Cascade Forward Neural Network (CFNN), Recurrent Neural Network (RNN), and Elman Neural Network (ENN)	NA	Robust, dependable and reliable algorithms Expensive Time-consuming	With the lowest RMSE, MSE, and MAE of 6.360, 40.45, and 0.54, respectively, the Elman NN has the greatest performance for the year 2018.
Shalini., et al. 2021 [22]	KNN, decision tree, random forest, linear regression	To collect real-time data from a test region, an IoT-based weather station has been developed.	Effective Good prediction No specific IoT device is mentioned.	Random forest: 0.083 Multiple linear regression: 0.165 Decision tree: 0.094 K-nearest neighbour: 0.103

(Continued)

TABLE 3.1 (Continued)
Comparison of Previous Work Done

Work done	ML method used	IoT device used	Pros and Cons	Accuracy/MAE
Syarifuddin et al. 2021 [28]	A geometry algorithm on an IoT platform to measure rainfall in real-time.	NA	The system uses an IoT platform, which allows for remote monitoring and data collection. Time delay	100% accuracy
Zhao et al. 2021 [23]	LR, Support Vector Classification, nearest neighbour, Gaussian Naïve Bayes, Bernoulli Naive Bayes, AdaBoost, Gradient Boosting, Bagging, random forest, Extra Trees, Gaussian Process Classification, logistic regression, Extra Tree, decision tree, Quadratic Discriminate Analysis.	Researchers used recordings of steady rainfall from five rainfall metres in a debris flow from December 2012 to April 2015; hence no IoT device is used.	The model is effective in predicting debris flow events based on rainfall data. Requires continuous rainfall monitoring	QDA – 100% ET – 99% AdaBoost – 98.5% RF – 98.5% XGB – 98.3% GB – 98% LR – 97.5% Bagging – 97% KNN – 96.7% SVC – 96% Nu-SVC -95.8% Gaussian NB – 88%
Sadhukhan et al. 2021 [24]	SVM, KNN DNN (Deep Neural Network) ANN (Artificial Neural Network)	GPS-enabled IoT devices like HC-12, DHT-11, BMP-180, BD-139 (transistor)	Accurate Portable Reliable System may not able to predict sudden changes	KNN – 70% LR – 84% SVM – 79.6% ANN – 89%
Rani et al. 2020 [25]	Linear regression, logistic regression, support vector machine, ANN	Water float sensors, rain drop sensors, IoT Geck	Easy to adapt Reliable Some-time prediction may not accurate	SVM – 90.60% ANN – 88%
Onkar et al 2019 [26]	Rainfall prediction using machine learning model (SVM)	Wireless sensor network (WSN) GPRS (General Pocket Radio Service) via a cellular network	Flexible Efficient detection of rainfall Time delay	100% Accuracy
Mzyece et al. 2018 [27]	Artificial Neural Network (ANN)	NA	Can handle more data at one time Complex and difficult to interpret.	ANN – 99% Best validation using ANN – 91% at epoch 44
Salmayenti et al 2017 [29]	Artificial Neural Network (ANN) for rainfall prediction	NA	High accuracy Accuracy decreases when used for long-term predictions	High prediction accuracy for monsoonal regions (R2: 0.59–0.82, RMSE: 0.04–0.09)

3.4 VISUALIZATION OF PREVIOUS WORK

This section of the chapter focuses on the comparative study of previous work done so that the reader can have a good idea of the percentage of accuracy achieved in this field. In order to present this section, some previous standard work has been compared in the earlier sections of the chapter.

Islam M. S. 2023 [30] has done a study for rain prediction. For that, various machine learning algorithms like Radial bias classifier, Multilayer perceptron, K-nearest neighbour, SVM, decision tree, logistic regression, CatBoost, LightGBM, Gaussian Naïve Bayes, AdaBoost, GRU, LSTM, and CNN have been applied; it is found that CNN with 100 epoch performs better in terms of accuracy. Figure 3.1 shows the comparison of the machine learning models in terms of accuracy. Islam M. S. 2023 [30] has outlined several strategies for enhancing the performance of rainfall prediction using AI and IoT. By leveraging XGBoost's ensemble learning abilities, they seek to enhance the model's precision in forecasting rainfall events. In this study, CatBoost and LightGBM, two ensemble models, are compared and both models consistently demonstrate higher accuracy than other models. This might indicate the strength of ensemble techniques in handling the complexities of the rainfall prediction task.

According to research by Islam et al. [31], Linear Gradient Boosting (LGB) with Select-K-Best feature selection exhibits the best performance in terms of regression, with test set scores of 0.203, MAE of 6.40, and RMSE of 15.44. Among the classifiers, XG-Boost (XGB) with no feature selection and no sampling has the greatest accuracy of 0.787 and an f1-score of 0.62 on the test set. With an R2-score of 0.189, an MAE of 5.789, and an RMSE of 15.575, the XGB classifier paired with LGB regression without any feature selection performed the best on the test set among the ZIR models. The ZIR models provide lower MAE scores and somewhat higher RMSE scores than regression models, despite being outperformed by them in terms of R2-scores. Figure 3.2 compares both the traditional LGB models and the XGB classifier/LGB regression combination.

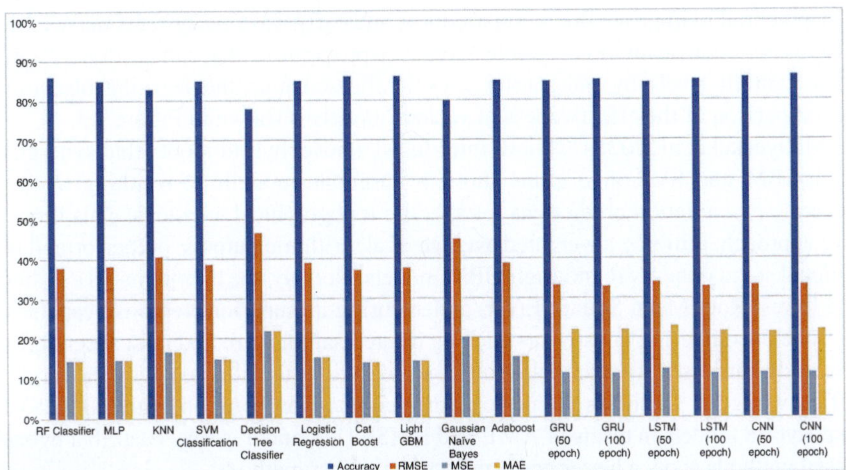

FIGURE 3.1 Performance comparison of prediction model [30].

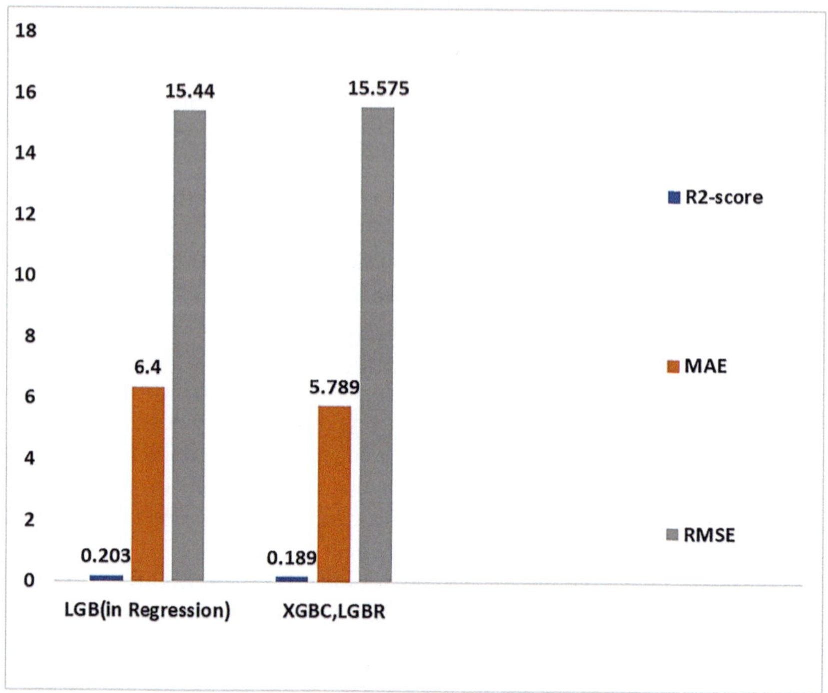

FIGURE 3.2 Comparison of LGB and combined XGBC/LGBR [31].

Lagrazon et al. 2023 [32] have provided a comparison analysis of machine learning models for performance evaluation of several machine learning models like Tree, SVM, Gaussian Process Regression, Ensemble, and Neural Network models. With the lowest RMSE and greatest R-squared values within this group of models, the Gaussian Process Regression model stood out as having the best overall performance. The potential to improve flood preparedness and early warning systems makes these discoveries significant. It is possible to create a programme that can accurately anticipate rainfall levels by integrating a well-validated machine learning algorithm. A comparison of the effectiveness of various models is shown in Figure 3.3.

Maliyeckel et al. 2021 [17] have introduced a novel hybrid model that synergizes LightGBM and SVR algorithms through ensemble modelling, resulting in more accurate precipitation predictions for specific topographical locations. This innovative approach, utilizing a weighted average of algorithmic outputs, outperformed traditional standalone SVR and LightGBM models. Notably, the hybrid model exhibited the lowest Root Mean Square Error, underscoring its superior predictive capability. Despite encountering constraints such as limited sample size and data accessibility, the study harnessed weather data from diverse stations to validate the hybrid model's efficacy. Figure 3.4 shows the performance comparison between LightGBM, SVR, and Hybrid models in terms of AME and RMSE [17], and it is very clear that hybrid-based methods show a better performance than other methods.

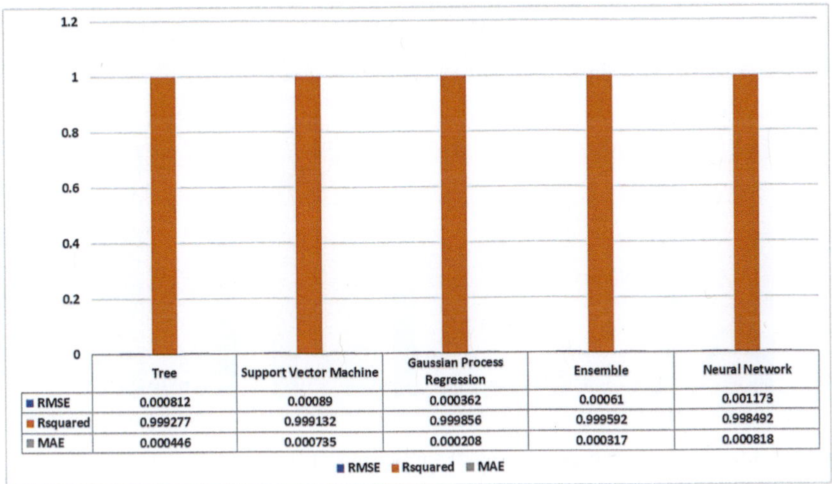

	Tree	Support Vector Machine	Gaussian Process Regression	Ensemble	Neural Network
RMSE	0.000812	0.00089	0.000362	0.00061	0.001173
Rsquared	0.999277	0.999132	0.999856	0.999592	0.998492
MAE	0.000446	0.000735	0.000208	0.000317	0.000818

FIGURE 3.3 Performance comparison for different models [32].

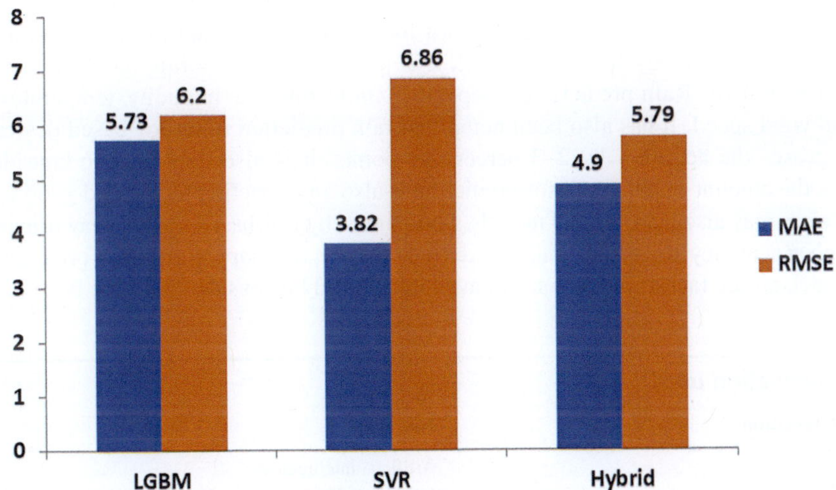

FIGURE 3.4 Performance comparison for LightGBM, SVR, and Hybrid method [17].

Figure 3.4 shows the performance comparison for lLightGBM, SVR, and Hybrid methods, and it shows the Mean Square Error for SVR. Figure 3.5 shows the performance comparison of various ML models for the 2023 and 2021 articles, and demonstrates that the highest prediction accuracy is achieved for the random forest method, then for linear regression, then for KNN, etc. The prediction accuracy is also highly dependent upon the data availability for training and testing, amount, and quality of data related to rain prediction.

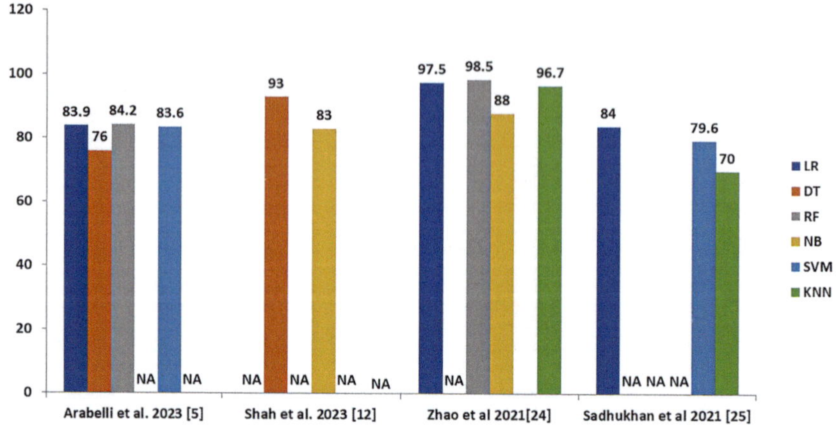

FIGURE 3.5   Performance comparison of various models [5], [12], [23], [24].

3.5   CONCLUSION

Rain prediction is an important concern for everyone, especially for farmers, and with new technology like IoT and AI, it is possible to predict rain more precisely and accurately. Rain prediction is dependent on factors like humidity, temperature, and wind speed. It has also been noted that rain prediction using IoT-based devices increases the accuracy by 2–3 percentage points. It is also important to consider that the amount of data used for prediction is also an impacting factor, and good IoT devices may also lead to good models. Sometimes, hybrid-based models may provide higher accuracy than traditional models. So before developing any rain-prediction model, all the factors and resources available should be considered for the best.

Abbreviation used

Abbreviation	Full word
AI	Artificial intelligence
IoT	Internet of Things
MLR	Multivariate linear regression
RF	Random forest
XG-Boost	Extreme gradient boost
LR	Logistic regression
ELSWP	Enhanced Learning Scheme for Weather Prediction
MAE	Mean Absolute Error
RMSE	Root Mean Square Error
SD	Standard deviation
KNN	K-nearest neighbour
ANN	Artificial Neural Network
ELM	Extreme learning machine
WKNN	Weighted K-nearest neighbour

Abbreviation used

Abbreviation	Full word
DWKNN	Dual weighted K-nearest neighbour
MEMS	Micro-electromechanical systems
ARMA	Autoregressive and moving average
SVM	Support vector machine
FFNN	Feed Forward Neural Network
CFNN	Cascade Forward Neural Network
RNN	Recurrent Neural Network
ENN	Elman Neural Network
DNN	Deep Neural Network

REFERENCES

1. Gupta, A., Mall, H. K., & Janarthanan, S. (2022). Rainfall Prediction Using Machine Learning. In 2022 First International Conference on Artificial Intelligence Trends and Pattern Recognition (ICAITPR) (pp. 1–5). Hyderabad, India. https://doi.org/10.1109/ICAITPR51569.2022.9844203.

2. Costa, V. G., & Pedreira, C. E. (2023). Recent Advances in Decision Trees: An Updated Survey. Artificial Intelligence Review, 56(5), 4765–4800.

3. Jaiswal, S., & Pandey, M. K. (2023). Digital Watermark Extraction Using RS-KNN and RS-LDA with LWT and Statistical Features. SN Computer Science, 4, 496. https://doi.org/10.1007/s42979-023-01924-9

4. Jaiswal, S., & Pandey, M.K. (2023). Deep Artificial Neural Network Based Blind Color Image Watermarking. In: Bhattacharyya, S., Banerjee, J. S., De, D., & Mahmud, M. (eds) Intelligent Human Centered Computing. Human 2023. Springer Tracts in Human-Centered Computing. Springer, Singapore. https://doi.org/10.1007/978-981-99-3478-2_10.

5. Arabelli, R., Sindhu, C., Keerthana, M., Madhav, T.V., Vamshi, C., & Ahmed, S.M. (2023). Improved Numerical Weather Prediction Using IoT and Machine Learning. In: Kumar, A., Mozar, S., & Haase, J. (eds) Advances in Cognitive Science and Communications. ICCCE 2023. Cognitive Science and Technology. Springer, Singapore. https://doi.org/10.1007/978-981-19-8086-2_109.

6. Rehman, A., Tariq, S., Farrakh, A., Ahmad, M., & Javeid, M. S. (2023). A Systematic Review of Machine Learning and Artificial Intelligence Methods to Tackle Climate Change Impacts. In 2023 International Conference on Business Analytics for Technology and Security (ICBATS) (pp. 1–7). Dubai, United Arab Emirates. https://doi.org/10.1109/ICBATS57792.2023.10111347.

7. Singh, S. K., Tiwari, A. K., & Paliwal, H. K. (2023). A State-of-the-art Review on the Utilization of Machine Learning in Nanofluids, Solar Energy Generation, and the Prognosis of Solar Power. Engineering Analysis with Boundary Elements, 155, 62–86.

8. Praveen, B., Talukdar, S., Shahfahad et al. (2020). Analyzing Trend and Forecasting of Rainfall Changes in India Using Non-parametrical and Machine Learning Approaches. Science Reports, 10, 10342. https://doi.org/10.1038/s41598-020-67228-7.

9. Nandgude, N., Singh, T. P., Nandgude, S., & Tiwari, M. (2023). Drought Prediction: A Comprehensive Review of Different Drought Prediction Models and Adopted Technologies. Sustainability, 15(15), 11684.

10. Dowlut, N., & Gobin-Rahimbux, B. (2023). Forecasting Resort Hotel Tourism Demand Using Deep Learning Techniques–A Systematic Literature Review. Heliyon, 2023(7), e18385,https://doi.org/10.1016/j.heliyon.2023.e18385.

11. Molina, M. J., O'Brien, T. A., Anderson, G., Ashfaq, M., Bennett, K. E., Collins, W. D., . . . & Ullrich, P. A. (2023). A Review of Recent and Emerging Machine Learning Applications for Climate Variability and Weather Phenomena. Artificial Intelligence for the Earth Systems, 1–46.

12. Shah, P., Khaitan, Y. A. U., & Kayalvizhi, S. (2023). Weather Management System Using Machine Learning Algorithm and IOT. In 2023 International Conference on Recent Advances in Electrical, Electronics, Ubiquitous Communication, and Computational Intelligence (RAEEUCCI) (pp. 1–4). Chennai, India. https://doi.org/10.1109/RAEEUCCI57140.2023.10134454.

13. Liyew, C. M., & Melese, H. A. (2021). Machine Learning Techniques to Predict Daily Rainfall Amount. Journal of Big Data, 8, 153. https://doi.org/10.1186/s40537-021-00545-4.

14. Sathya, P., & Gnanasekaran, P. (2023). Rainfall Forecasting System Using Machine Learning Technique and IoT Technology for a Localized Region. In: Shakya, S., Du, KL., & Ntalianis, K. (eds) Sentiment Analysis and Deep Learning. Advances in Intelligent Systems and Computing, vol. 1432. Springer, Singapore. https://doi.org/10.1007/978-981-19-5443-6_32.

15. Sharma, S. A., Gangopadhyay, A., Koushik, K. T., Sriharipriya, K. C., & Clement, J. C. (2022). Real-Time Rainfall Prediction System Using IoT and Machine Learning. In: Arunachalam, V., & Sivasankaran, K. (eds) Microelectronic Devices, Circuits and Systems. ICMDCS 2022. Communications in Computer and Information Science, vol 1743. Springer, Cham. https://doi.org/10.1007/978-3-031-23973-1_10

16. Xu, Y., Hu, C., Wu, Q., Jian, S., Li, Z., Chen, Y., Zhang, G., Zhang, Z., & Wang, S. (2022). Research on Particle Swarm Optimization in LSTM Neural Networks for Rainfall-runoff Simulation. Journal of Hydrology, 608, 127553. https://doi.org/10.1016/j.jhydrol.2022.127553

17. Maliyeckel, M. B., Sai, B. C., & Naveen, J. (2021). A Comparative Study of LGBM-SVR Hybrid Machine Learning Model for Rainfall Prediction. In 2021 12th International Conference on Computing Communication and Networking Technologies (ICCCNT) (pp. 1–7). Kharagpur, India. https://doi.org/10.1109/ICCCNT51525.2021.9579628.

18. Dash, Y., Mishra, S. K., & Panigrahi, B. K. (2018). Rainfall Prediction for the Kerala State of India Using Artificial Intelligence Approaches. Computers and Electrical Engineering, 70. https://doi.org/10.1016/j.compeleceng.2018.06.004.

19. Huang, M., Lin, R., Huang, S., & Xing, T. (2017). A Novel Approach for Precipitation Forecast via Improved K-nearest Neighbor Algorithm. Advances in Engineering Informatics, 33, 89–95. https://doi.org/10.1016/j.aei.2017.05.003.

20. Chao, Z., Pu, F., Yin, Y., Han, B., & Chen, X. (2018). Research on Real-Time Local Rainfall Prediction Based on MEMS Sensors. Journal of Sensors, 2018, e6184713. https://doi.org/10.1155/2018/6184713.

21. Dada, E.G., Yakubu, H.J., & Oyewola, D.O. (2021). Artificial Neural Network Models for Rainfall Prediction. European Journal of Electrical Engineering and Computer Science, 5(2), 30–35, https://doi.org/10.24018/ejece.2021.5.2.313.

22. Shalini, H., & Aravinda, C.V. (2021). An IoT-Based Predictive Analytics for Estimation of Rainfall for Irrigation. In: Chiplunkar, N.N., & Fukao, T. (eds) Advances in Artificial Intelligence and Data Engineering. AIDE 2019. Advances in Intelligent Systems and Computing, vol 1133. Springer, Singapore. https://doi.org/10.1007/978-981-15-3514-7_105

23. Yan, Z., Xingmin, M., Tianjun, Q., Yajun, L., Guan, C., Dongxia, Y., Feng, Q. (2021). AI-based Rainfall Prediction Model for Debris Flows. Engineering Geology, 106456. https://doi.org/10.1016/J.ENGGEO.2021.106456

24. Sadhukhan, M., Dasgupta, S., & Bhattacharya, I. (2021). An Intelligent Weather Prediction System Based on IOT 2021 Devices for Integrated Circuit (DevIC) (pp. 528–532), Kalyani, India. https://doi.org/10.1109/DevIC50843.2021.9455883.

25. Rani, D. S., Jayalakshmi, G. N., & Baligar, V. P. (2020). Low Cost IoT based Flood Monitoring System Using Machine Learning and Neural Networks: Flood Alerting and Rainfall Prediction. 2020 2nd International Conference on Innovative Mechanisms for Industry Applications (ICIMIA). https://doi.org/10.1109/icimia48430.2020.9074.
26. Amale, O., & Patil, R. (2019). IOT Based Rainfall Monitoring System Using WSN Enabled Architecture. In 2019 3rd International Conference on Computing Methodologies and Communication (ICCMC) (pp. 789–791). Erode, India. https://doi.org/10.1109/ICCMC.2019.8819721.
27. Mzyece, L., Nyirenda, M., Kabemba, M. K., & Chibawe, G. (2018). Forecasting Seasonal Rainfall in Zambia – An Artificial Neural Network Approach. Zambia ICT Journal, 2(1), 16–24. https://doi.org/10.33260/zictjournal.v2i1.46.
28. Syarifuddin, S. D. S., Khurniawan, A., Aulia, S., Ramadan, D. N., & Hadiyoso, S. (2021). Rainfall Information System Using Geometry Algorithm on IoT Platform. In 2021 IEEE Asia Pacific Conference on Wireless and Mobile (APWiMob) (pp. 195–199). Bandung, Indonesia. doi: 10.1109/APWiMob51111.2021.9435219.
29. Salmayenti, R., Hidayat, R. & Pramudia, A. (2017). Rainfall Prediction Using Artificial Neural Network. Journal Agromet Indonesia, 31(1), 11–21. https://doi.org/10.29244/J.AGROMET.31.1.11-21.
30. Islam, M. S., Hossain, A., Khatun, A., & Kor, A. (2023). Evaluation of the Performance of Machine Learning and Deep Learning Techniques for Predicting Rainfall: An Illustrative Case Study from Australia. https://doi.org/10.1109/ECCE57851.2023.10101560.
31. Islam, F. Z., Islam, R., Jamil, A., & Momen, S. (2022). Evaluation of Machine Learning Methods for Predicting Rainfall in Bangladesh. https://doi.org/10.1109/CITDS54976.2022.9914171.
32. Lagrazon, P. G. G. et al., (2023). A Comparative Analysis of the Machine Learning Model for Rainfall Prediction in Cavite Province, Philippines. IEEE World AI IoT Congress (AIIoT) (pp. 0421–0426), Seattle, WA. https://doi.org/10.1109/AIIoT58121.2023.10174533.

4 Machine Learning-Based Prediction of Wind Speed for Ratnagiri Region, India

Nabanita Mandal and Sarode Tanuja

4.1 INTRODUCTION

Climate prediction for a region is a challenging task. The temperature, rainfall, humidity, and wind speed of that region are related to several factors. In Maharashtra, the Ratnagiri District of the Konkan coast lies along the Arabian Sea. This region receives a good amount of rainfall during the monsoons because of the winds that blow from the southwest direction. The special phenomenon for this region is the presence of very strong winds leading to cyclones before and after the monsoons. Prediction of wind speed in any region helps to identify the potential for power generation through wind energy. Accurate predictions are important for planning renewable energy projects. In this research, an attempt has been made to predict the wind speed using Machine Learning techniques. These data-driven techniques learn from the patterns created by data collected over the years. It understands the variations in the parameters over time and uses them to predict future values. The Long Short-Term Memory-based proposed model, random forest model, and Support Vector Regression techniques are implemented for the data to get a suitable prediction. The results presented graphically are analyzed. This chapter consists of the following sections: Literature Survey, Data Set, Techniques, Results, Conclusion, Acknowledgment, and References.

4.2 LITERATURE SURVEY

Soukayna M., et al. have researched exploring the ability of neural network techniques [1]. These were used for downscaling of one climate parameter to a given region of interest. A new statistical downscaling model was proposed by Saptarshi Misra et al. on Recurrent Neural Networks. LSTM was implemented, which captures the dependencies in the data used for predicting rainfall locally [2]. Bochenek et al. analyzed different models for weather prediction [3]. Ju-Young Shin et al. have proposed a model based on random forests for forecasting the speed of wind [4]. In the paper by Senthil Kumar P, the same has been discussed using different techniques with Artificial Neural Networks [5]. For dealing with the change in the climate, machine

DOI: 10.1201/9781003452393-4

learning can be utilized to give significantly good results. This has been published in research by David Rolnick et al [6]. Jude Chukwura Obi has compared multiple regressions and Support Vector Regressions in his research using 15 different datasets. The findings show that Root Mean Square Error is lower in multiple regression than in Support Vector Regression [7]. Considering the importance of using Machine Learning Techniques on different climate parameters, the prediction of wind speed on monthly data has been implemented by using Multivariate Regression, Deep Neural Networks, and random forests [8]. For the prediction of climate parameters throughout a region, it is necessary to understand the trends. The prediction is done considering the change in those climate parameters over the years. Comparison of Artificial Neural Network, ridge regression, and polynomial regression helps with the accuracy. This has been researched by Ferdous et al [9]. Jian He and Jingle Xu have developed a prediction model with improved accuracy by support vector machine, which uses a function of the combined kernel [10]. Lili Wang et al. have collected data from many locations and created a model using an extreme learning machine and AdaBoost, which they have used for predicting the speed of wind at different times [11]. Yun Zheng et al. have suggested a ridge regression model for wind speed forecasting [12]. The uncertainty in the wind speed pattern leads to reduced accuracy. To solve this problem, Support Vector Regression has also been used [13]. Koffi et al. have used Support Vector Regression for mean wind speed hourly prediction [14].

4.3   DATASET

The data for the Ratnagiri Region is collected from the India Meteorological Department, Pune, Maharashtra, India [15]. It contains weekly surface data from 2010 onwards. It consists of various parameters relevant for wind speed (WSP) prediction, like relative humidity (WRH), vapor pressure (WVP), wet bulb temperature (WWBT), dry bulb temperature (WDBT), and dew point temperature (WDPT) which are measured for two synoptic hours 3 and 12. The description of the parameters is as follows:

- WSP: It is the rate of airflow measured in meters per second, kilometers per hour, or miles per hour. The atmospheric pressure, temperature, and rotation of the earth are some of the factors that influence the wind speed. The intensity and strength of the wind speed are major factors that affect aviation, wind power generation, and sailing.
- WRH: It is the amount of water vapor present in the air in comparison with maximum capacity at a specific temperature. It is expressed in a percentage. It is an important factor that contributes to cloud formation, precipitation, storms, and drought. The agriculture sector is affected by the humidity levels because the amount of water required by plants and the moisture present in the soil determines the quality of crops.
- WVP: The partial pressure of the water vapor is the vapor pressure. It helps in determining the humidity, precipitation, and temperature that contribute to extreme weather events.

- WDBT: Dry Bulb Temperature is also known as the air temperature. It is the temperature without considering the moisture content present in the air. It can influence the type of precipitation, like rain or snow. It is highly significant in the design of ventilation and air conditioning systems.
- WWBT: The lowest temperature to which the air can be cooled at a constant pressure by evaporation of water is the Wet Bulb Temperature. It is an important parameter to calculate other meteorological parameters like dew point temperature, heat index, humidity, etc. It is important for understanding the condition of the environment and the availability of water.
- WDPT: The temperature at which air becomes saturated with moisture is the dew point temperature. At constant pressure, the air must be cooled at a certain temperature. That temperature is considered as the dew point temperature. It helps in predicting the formation of clouds, the likelihood of precipitation, and relative humidity.

4.3.1 Data Pre-Processing

Before applying Machine Learning or Deep Learning algorithms, it is important to clean the data. Since the results of these algorithms heavily rely on huge data, it is critical to have a clear understanding of the data. For this, there are various data visualization tools and techniques available that help in the graphical representation of the data. These data visualization tools help provide insights about the data. This helps us understand the nature of the data. Figure 4.1 represents the Wind Speed weekly data from 2010. It is a line chart that depicts the trends over time.

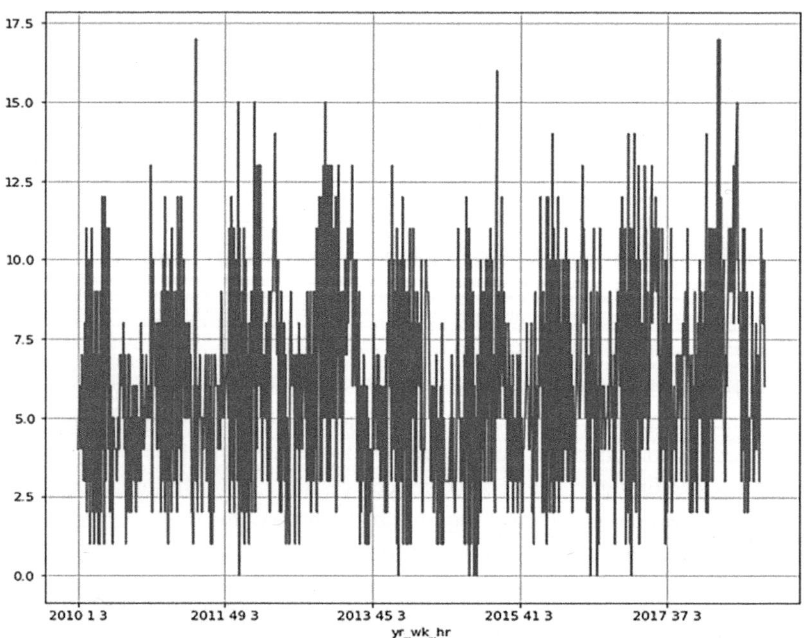

FIGURE 4.1   Plot of WSP.

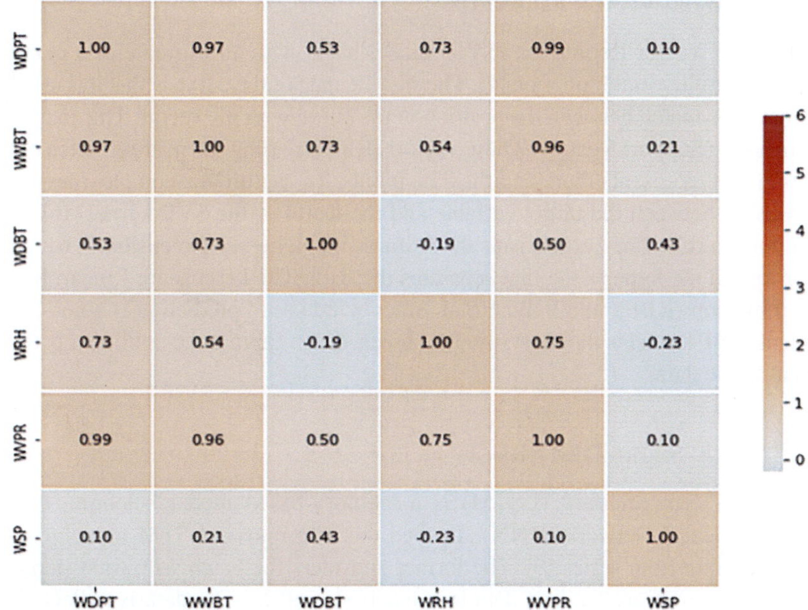

FIGURE 4.2   Heatmap to show correlation.

From the figure, it can be observed that there exist seasonal variations. There are a lot of factors that influence the wind speed of a region. How these factors correlate with each other is depicted in V.

The heatmap represents the correlation between all variables. The dark colors are for high correlation, and the light colors are for low correlation.

4.3.2 DATA ARRANGEMENT

After data preprocessing is done, the next step is to arrange the data so that it can be supplied to the model. At this point, the complete data set is converted to two disjointed datasets. One is used in training, and the other one is used in testing. A larger percentage of data is to be utilized as a training sample, and less data is considered in testing. This is because the model learns from the training and uses testing for prediction. In this research, the two approaches are used and the results are analyzed. In the first approach, 90% of data is applied to train the model, and 10% is kept for the testing. The second approach uses 80% of the entire data to perform training, and for testing 20% data is used. The target variable here is WSP, and the features are WWBT, WDBT, WDPT, WRH, and WVP.

4.4 TECHNIQUES APPLIED

This section describes the machine learning techniques that are implemented in the data set. Support Vector Regression (SVR), Long Short-Term Memory (LSTM), and the Proposed Model are implemented on the climate data to get the prediction of wind speed.

4.4.1 SUPPORT VECTOR REGRESSION

The Support Vector Regression (SVR) model helps with an estimate of a continuous-valued multivariate function [7]. The significance of the curve is that it is used for obtaining the match between the position of the curve with the vector. This matching is done using Support Vectors. These vectors help in finding the nearest match of the function that represents them and the data points. To capture the complexities of the relationship between the target variable and the features, the SVR kernel [16] plays an important role. The kernel maps the features into a higher dimension. It becomes easier to find the hyperplane that separates the data. The kernels are Linear, Radial Basis Function (RBF), and Polynomial. Since wind speed prediction is a non-linear problem, RBF kernel is the most suitable choice. It can capture the nonlinear patterns present in the data.

4.4.2 LONG SHORT-TERM MEMORY

Long Short-Term Memory (LSTM) is a memory-based model belonging to the Recurrent Neural Network (RNN). To feed the data into an LSTM model, data is rearranged into time series [9]. The format considers the batch size, time step, and features. Here, we are feeding data in a batch size of 32. The data is collected for two synoptic hours; hence we can have a lookback of two, which will be used as the time step. The LSTM model uses the input gate to take the input values, then decides which values need not be remembered using the forget gate and generates the output using the output gate [17]. It consists of a single-layer LSTM with separate input and output layers. To test this model again, we are using both approaches for the training testing split. The learning rate is maintained at 0.01 for the model. The learning rate is a hyperparameter that decides how fast or slow the model will learn.

4.4.3 PROPOSED MODEL

The proposed model architecture is based on LSTM. It consists of one input layer, one output layer, and three hidden layers. In these hidden layers, the first is the LSTM layer, which helps capture the dependencies in the sequential data. The second and third hidden layers are fully connected dense layers. These two layers help transform and capture the features extracted by the LSTM layer. All three layers work in a hierarchy to learn and understand the hidden patterns and relationships within the input data.

4.5 RESULT ANALYSIS

This section describes the plots obtained after prediction for the three different approaches of SVR, LSTM, and Proposed Model.

Figure 4.3 shows the actual and predicted values of the target variable, WSP, when the data is trained for 80% and tested for 20% using SVR.

When the SVR model is implemented, consider 90% for training and 10% for testing. In Figure 4.4, we can see that there is a variation in the results compared to the previous graph.

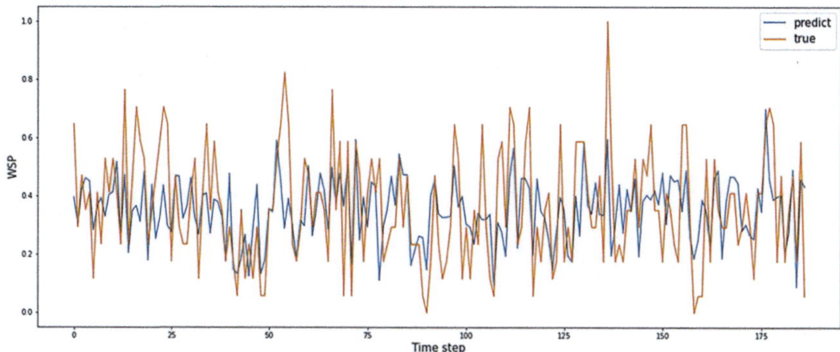

FIGURE 4.3 Actual vs predicted WSP for 80:20 for SVR.

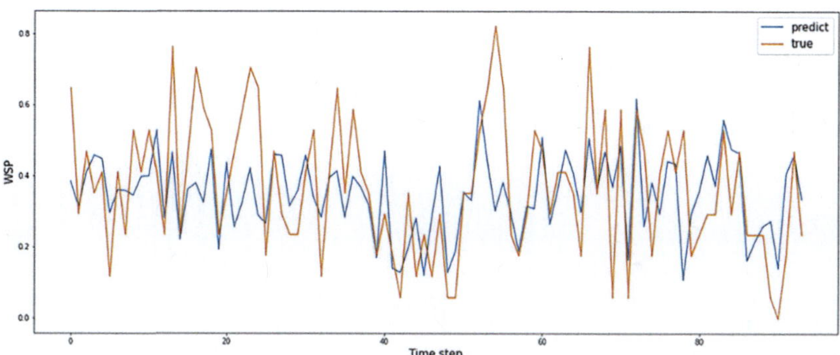

FIGURE 4.4 Actual vs predicted WSP for 90:10 for SVR.

Here it is observed that Support Vector Regression gives predictions that are not as close to the actual ones. From the graphs in Figure 4.3 and Figure 4.4, we can say that the trend is properly captured by SVR but lacks accuracy. The performance of SVR is affected due to improper regularization. The model can't handle the complexity of the data properly. To solve these problems, a model is required that improves accuracy. So, Long Short-Term Memory (LSTM) is applied here.

Figure 4.5 shows the output of actual WSP and predicted WSP. The LSTM model captures the pattern quite closely.

Now the training testing split is changed to 90:10.

In Figure 4.6, the prediction of WSP using LSTM is shown with the training testing split of 90:10.

There is still scope for improving the predictions. So, the proposed model is applied here.

Figure 4.7 represents the actual and predicted WSP for the training testing split of 80:20.

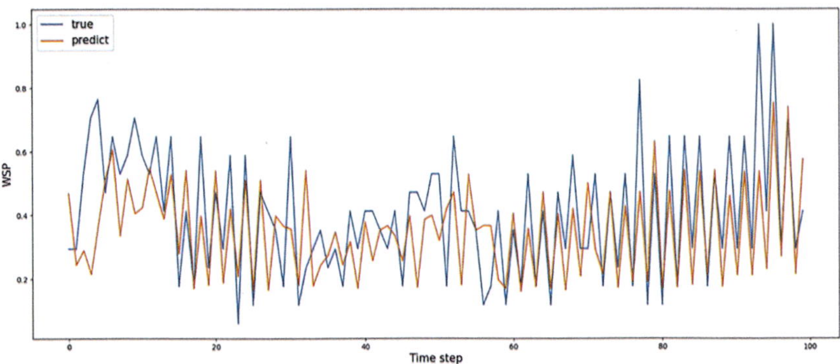

FIGURE 4.5 LSTM output for WSP with 80:20.

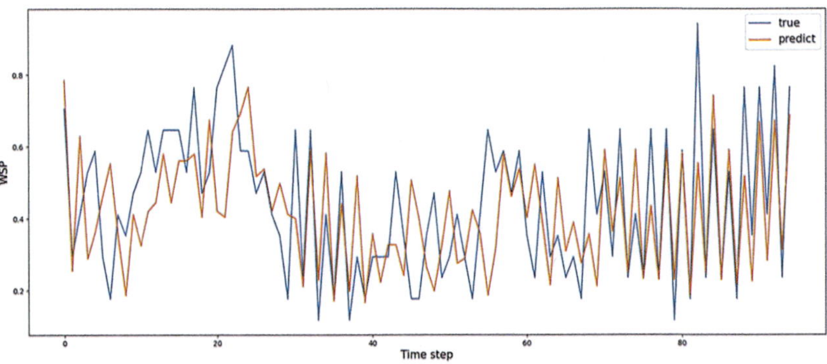

FIGURE 4.6 LSTM output for WSP with 90:10.

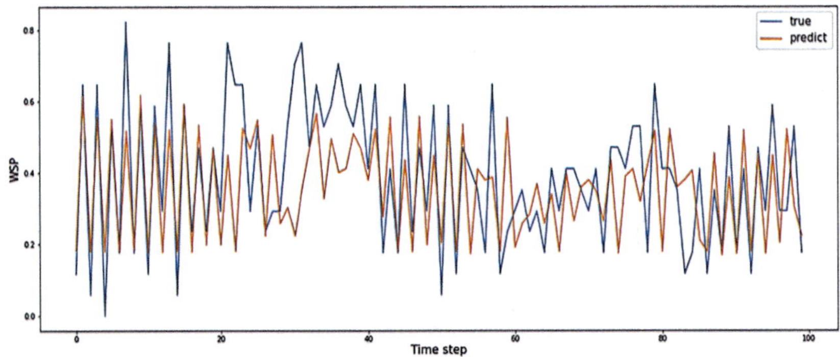

FIGURE 4.7 Proposed model output for WSP with 80:20.

The Proposed Model is applied to a training and testing split of 90:10 and it is observed that the predictions are improved. Figure 4.8 shows that the predicted and actual values are closer.

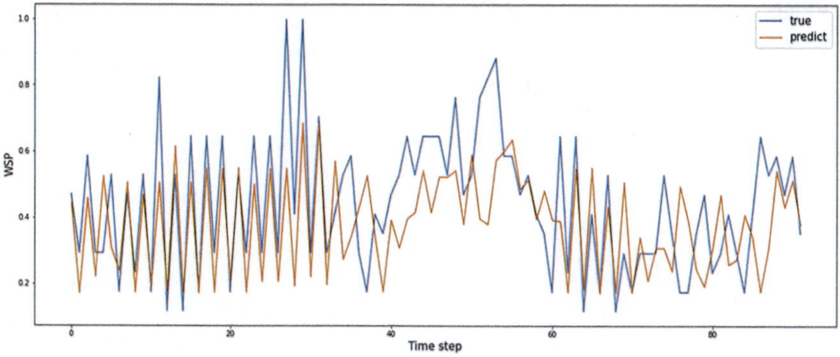

FIGURE 4.8 Proposed model output for WSP with 90:10.

TABLE 4.1
Comparison of SVR, LSTM, and Proposed Model

Train-Test Split	Proposed Model	LSTM	SVR
90%:10%	0.127204	0.130282	0.438797
80%:20%	0.130267	0.131666	1.394696

Table 4.1 represents the comparison between Mean Absolute Error [18] values for SVR, LSTM, and the Proposed Model for both training testing splits of 90:10 and 80:20.

For both approaches, the error obtained is lower in the Proposed Model than the LSTM and SVR models.

4.6 CONCLUSION

After implementing SVR, LSTM, and the Proposed Model, it can be inferred from the plots that the Proposed Model gives better prediction results. SVR can attain the trend present in the data. The SVR model learns from the data and shows the prediction, but it is not as accurate. LSTM on the other hand captures the trend properly and gives a lower prediction error. Among all these models, the error is reduced when the training-testing split is 90:10. This is because more data is considered for training, so the model can better understand the hidden patterns and give more accurate predictions. Prediction of wind speed for the Ratnagiri region is a significant task because it decides the humidity, storms, cyclones, and temperatures of the region.

ACKNOWLEDGMENTS

The procurement of data is done by the India Meteorological Department, Pune, Maharashtra, India. We are indebted for enabling this research.

REFERENCES

1. Mouatadid, S., Easterbrook, S., & Erler, A. A machine learning approach to non-uniform spatial downscaling of climate variables. In IEEE International Conference on Data Mining Workshops (ICDMW), 332–341 (2017).
2. Misra, S., Sarkar, S. & Mitra, P. Statistical downscaling of precipitation using long short-term memory recurrent neural networks. Theor Appl Climatol, 1179–1196 (2018).
3. Bochenek, B., & Ustrnul, Z. Machine learning in weather prediction and climate analyses—applications and perspectives. Atmosphere, 13(2), 180 (2022).
4. Shin, J. Y., Min, B., & Kim, K. R. High-resolution wind speed forecast system coupling numerical weather prediction and machine learning for agricultural studies—a case study from South Korea. Int J Biometeorol, 66, 1429–1443 (2022).
5. Paramasivan, S. Improved prediction of wind speed using machine learning. EAI Endors Trans Energy Web, 6 (2019).
6. David, R., Priya, L., et al. Tackling climate change with machine learning. ACM Comput Surv, 55(2), 42 (2022).
7. Obi, J. C. A comparative study of the multiple regression and support vector regression. COOU J Phys Sci, 3(1), 465–473 (2020).
8. Mandal, N., & Sarode, T. Prediction of wind speed using machine learning. Int J Comput Appl, 176(32), 34–37 (2020).
9. Antor, A. F., & Wollega, E. Comparison of machine learning algorithms for wind speed prediction. In Proceedings of the 5th NA International Conference on Industrial Engineering and Operations Management, Detroit, MI, August 10–14 (2020).
10. He, J., & Xu, J. Ultra-short-term wind speed forecasting based on support vector machine with combined kernel function and similar data. J Wireless Comput Netw, 2019, 248 (2019).
11. Wang, L., Guo, Y., Fan, M., & Li, X. Wind speed prediction using measurements from neighboring locations and combining the extreme learning machine and the AdaBoost algorithm. Energy Rep, 8 (2022).
12. Zheng, Y., Ge, Y., Muhsen, S., Wang, S., Elkamchouchi, D. H., Ali, E., Elhosiny Ali, H. New ridge regression, artificial neural networks and support vector machine for wind speed prediction. Adv Eng Softw, 179 (2023).
13. Fu, X., Feng, Z., Yao, X., & Liu, W. A novel twin support vector regression model for wind speed time-series interval prediction. Energies, 16 (2023).
14. Agbeblewu Dotche, K., Salami, A. A., Kodjo, K. M., Ouro-Agbake, H., & Koffi-Sa, B. Wind speed prediction based on support vector regression method: a case study of Lome-Site. In 2019 IEEE PES/IAS Power Africa, Abuja, Nigeria, 267–272 (2019).
15. India Meteorological Department, www.imdpune.gov.in/
16. Bakar, A., Aftar, M., Ariff, M., Noratiqah, Nadzir, M., Shahrul, M., & Lau, Y. Comparison of support vector regression (SVR) kernel functions for predicting PM10 time series data in Malaysia. In Proceedings of the 6th International Conference on Mathematical Applications in Engineering, January (2023).
17. Marndi, A., & Patra, G. K. Multidimensional ensemble LSTM for wind speed prediction. In Sharma, H., Gupta, M. K., Tomar, G. S., & Lipo, W. (eds) Communication and Intelligent Systems. Lecture Notes in Networks and Systems, Vol. 204, Springer, Singapore (2021).
18. Hodson, T. O. Root mean square error (RMSE) or mean absolute error (MAE): when to use them or not. Geosci Model Develop, 15, 5481–5487 (2022).

5 Wind Power Forecasting with Machine Learning Approach

Rakesh Kumar, Prakash M and Shakila B

5.1 INTRODUCTION

Wind energy is a well-known kinetic energy resulting from a voluminous shift of air. The displacement of wind is caused due to uneven heating of atmosphere by the sun, thus creating differences in temperature, density and pressure [1]. The existence of wind can be felt all around us, and with significant power density in some areas. Wind energy has been utilised for centuries to pump water, sailboats, grind grain, etc.

The discovery of fossil fuels had led to a shift in their use. The use of nonrenewable power sources has significant environmental impacts such as air pollution and global warming. The nations, energy firms and individuals are all paying more attention to renewable energy sources since they are readily available in nature, cost-free and environmentally friendly. Wind power is one of the most significant and promising renewable energy sources [2].

The global wind energy market is experiencing significant growth among all renewable sources due to its ample availability and low-cost energy production. Several countries have acknowledged wind power as the best suitable alternate in future power generation. Subsequently, the installed wind capacity has gradually increased more than 30% annually [3]. According to the World Energy Review in 2021, wind energy has supplied more than 1800 *TWh* of electricity, which was over 6% of the world's electricity demand and around 2% of world energy with added 100 GW.

Based on the Global Wind Energy Council (GWEC) report, the global capacity of wind power installation increased from approximately 190 GW in 2010 to more than 740 GW in 2020. As per the 2021 Renewable Capacity Statistics provided by the International Renewable Energy Agency, a 14.3% increase from the previous year has been observed in wind power installed capacity. The top 10 countries' total wind capacity in 2020 are shown in Figure 5.1 [4]. Globally, nearly 56% of total wind energy is produced by China and the United States of America.

India has around 39 GW wind power-installed capacity and comes in fourth place. It is observed that China has the biggest proportion, with 218.991 GW of worldwide capacity, followed by the United States with 117.774 GW. Overall, the installed WP capacity has increased significantly over the last 11 years. Integrating wind power

DOI: 10.1201/9781003452393-5

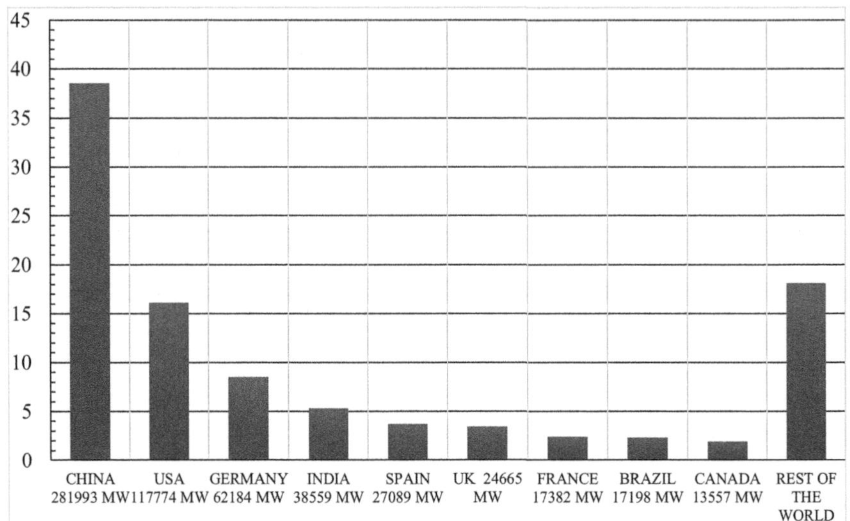

FIGURE 5.1   Global wind power install capacity.

with existing electrical grids will reduce pollution while satisfying the load demand. However, wind is an erratic source of energy that has a significant impact on the stability of the integrated grid system. The main difficulties with wind power are scheduling, management and optimization.

Forecasting of wind power plays a vital role in the power industry. Accurate and consistent wind power forecasting can balance and integrate various volatile power sources at all levels of transmission and distribution grids [5]. Furthermore, reliable short-term wind power forecasting can lessen issues brought on by grid integration of renewables and energy trade. Forecasting also plays a crucial role in reducing operating costs and increasing wind power's competitiveness [3, 6, 7, 9]. Understanding the many parameters that affect wind power is the first step in creating a machine learning model for forecasting. It is influenced by a number of factors, including temperature, pressure, wind direction, wind speed, etc.

5.2   WIND POWER FORECASTING

Wind is moving air that has mass. The kinetic energy of moving air can be given by equation (1), where v is velocity and m is mass flowing through the air A and density ρ. Equation (2) shows the propositional context of power generated by wind speed flowing through area A and air density ρ.

$$P_k = \frac{1}{2}mv^2 \tag{1}$$

$$P_k = \frac{1}{2}\rho A v^3 \tag{2}$$

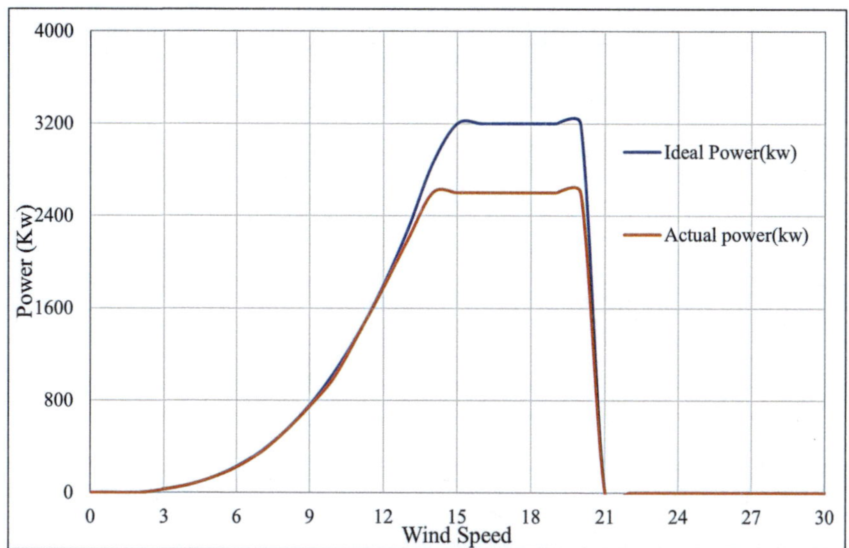

FIGURE 5.2   Power curve of wind turbine.

To extract the kinetic energy of wind, we need to convert it into usable form. Wind energy can be transformed into electrical power by a wind turbine. The Betz-limit, which represents the highest value for the power coefficient, states that a wind turbine can only collect 59% of the power present in the wind [10]. The curvilinear relation between wind speed and turbine power production is known as the power curve of turbine. Figure 5.2 shows a typical wind power curve of an S133 turbine manufactured by SUZLON ENERGY Ltd. It shows that the cut in and cut out speed of the turbine is 3 m/s and 20 m/s, respectively. In ideal conditions, the power output of the turbine is 3.2 MW at rated wind speed. But practically, it is only 2.6 MW.

The equation for wind power generation is given as

$$P_t(v) = \begin{cases} 0, & v < v_{ci} \\ (a_n v^n + a_{n-1} v^{n-1} + \ldots a_1 v^1 + a_0), & v_{ci} \le v < v_r \\ P_r, & v_c \le v < v_{co} \\ 0, & v \ge v_{co} \end{cases} \qquad (3)$$

Where $P_t(v)$ is power output, v_{ci} is cut in speed of wind, v_{co} is cut out speed, v_r is rated wind speed and P_r is rated output power of wind turbine.

5.3   MACHINE LEARNING ALGORITHMS

Machine learning is the science of teaching computers to learn by feeding sets of input and output along with the relationship between them [11]. Data input, abstraction and generalisation are the three processes that make up the machine learning process. Data input involves using recorded or observed data; abstraction enlarges

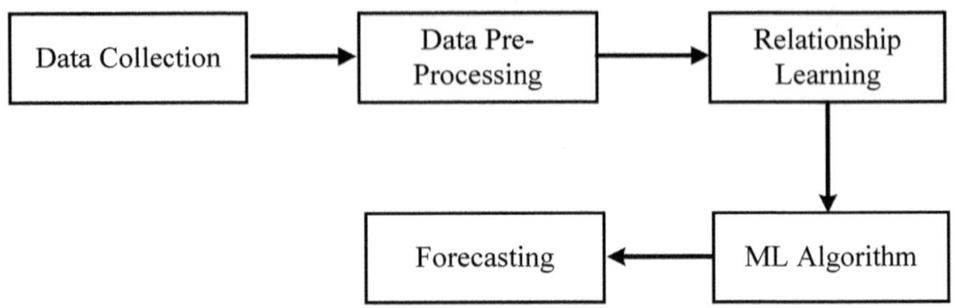

FIGURE 5.3 Wind speed vs wind power forecasting flowchart.

the data's scope; and generalisation affects how the abstracted data are used [12]. Machine learning algorithms also work when data features are not directly related to output variables. It is used for describing the behaviour of the dataset, input features with respect to expected output, etc.

Machine learning algorithms can be classified into three categories: Supervised learning, Unsupervised learning and Reinforcement. This study focuses on supervised machine learning algorithms, and the output is predicted using well labelled training datasets. Supervised learning is further classified into two categories: Classification and Regression. Classification models deal with categorical output value, and Regression model are often used for continuous output value [13, 14]. The sequence of processes used for forecasting the wind speed/wind power is shown in Figure 5.3.

5.3.1 IMPLEMENTATION OF LINEAR REGRESSION ALGORITHM

One of the simplest and most widely used machine learning techniques is linear regression (RL), which is very useful for enormous datasets [15]. It shows the relationship between the input feature and the output variable. The linear regression model provides a sloped straight line which best fits the output and input features [16]. The equation (4) represents the regression line for single input features, where y is output, x is an input feature, w is the weight of the input feature x and b is the intersection point.

$$y = wx + b \tag{4}$$

$$y = \theta_0 x_0 + \theta_1 x_1 + \theta_2 x_2 \cdots\cdots + \theta_n x_n \tag{5}$$

The equation (5) illustrates a regression line that has n input features. Here, x_i is the i_{th} input feature and θ_i is the weight of that input feature. The varied values of weights ($\theta 1,2, \theta 3. \ldots \ldots .\theta_n$) result in different lines of regression. To determine the best fit line, the optimal input feature weight has to be derived by computing the loss function and gradient descent. Loss function describes the error in predicted value of output feature [17]. The loss function is given by

$$J(\theta_0 + \theta_1 + \cdots \theta_n) = \frac{1}{m} \sum_{i=1}^{m} (y_{i\,predicted} - y_{i\,actual})^2 \tag{6}$$

Where m is number of rows in dataset, $y_{i\,predicted}$ is predicted output by linear regression model and $y_{i\,actual}$ is the actual output feature of dataset. The minimisation of loss function can optimize the linear regression model, and it can be achieved with gradient descent. Gradient descent is an iterative optimisation technique used for finding minima of any differentiable function [18]. It is given by the partial derivative of the loss function with respect to the weight of the feature θ_i

$$\text{Gradient} = \frac{\partial \text{loss. function}}{\partial \theta} \tag{7}$$

For n = 0, and $x_0 = 1$

$$\text{Gradient} = \frac{1}{2m} \sum_{j=1}^{m} (y_{i\,predicted} - y_{i\,actual}) \tag{8}$$

For n > 1

$$\text{Gradient} = \frac{1}{2m} \sum_{j=1}^{m} (y_{i\,predicted} - y_{i\,actual}) * x_i^{(j)} \tag{9}$$

Gradient descent is an iterative technique that begins with a random value of $J(\theta_0, \ldots \theta_n)$ and then updates it in every iteration [19]. If the starting value is θ_j, then its gradient is computed and multiplied with learning rate θ before updating the result. For n = 0, equation for modification of θ_0 is given as

$$\theta_0 = \theta_0 - \alpha * \frac{1}{2m} \sum_{j=1}^{m} (y_{i\,predicted} - y_{i\,actual}) \tag{10}$$

And for n > 1,

$$\theta_j = \theta_j - \alpha * \frac{1}{2m} \sum_{j=1}^{m} (y_{i\,predicted} - y_{i\,actual}) * x_i^{(j)} \tag{11}$$

Where is feature i with *jth* training example. This modification process repeats till loss function converges. The significance of learning rate (α) is that smaller α will take more time (slow convergence) and gives minimum errors, whereas large α computes faster but it may not converge. In this chapter, the assumed value of α is 0.001.

5.3.1.1   Implementation of Decision Tree Regression Algorithm

Decision tree regression (DTR) is a non-parametric supervised machine learning algorithm used for continuous or numeric datasets [20]. It consists of root node, internal node and leaf node. The starting point of every decision tree is called root node. The node where decisions are made for splitting branches are called internal nodes, and the nodes which are not further divided are leaf nodes. The results are taken from leaf node only. Decision tree regression's hierarchical structure is built by splitting features and dividing the data domain into subdomains, then mapping the subdomain with output responses that meet maximum variance reduction criteria [21].

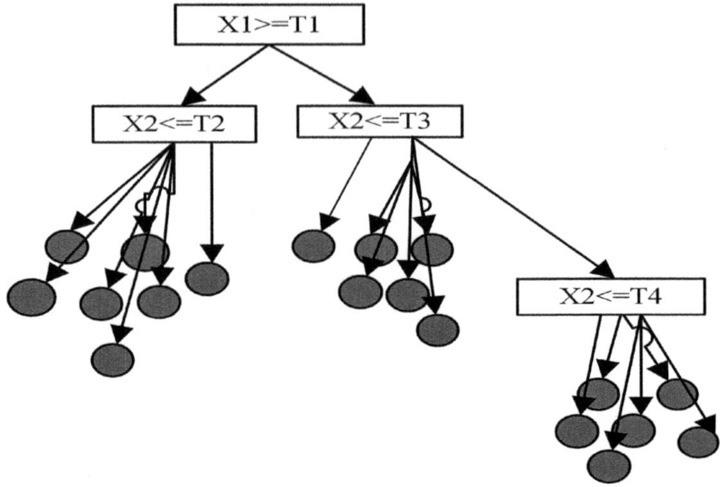

FIGURE 5.4 Decision Tree

Variance is the mean of the square difference of the output feature and the mean of the all-output feature, and it is given as

$$\text{Variance} = \frac{1}{n}\sum\nolimits_{i=1}^{n}\left(y_{i\,output} - y_{i\,mean}\right)^{2} \tag{12}$$

Where $y_{i\,output}$ is predicted output and $y_{i\,mean}$ is mean of output feature.
  The Variance Reduction (VR) is given as

$$VR = variance(parent) - \sum Wi * Variance(child) \tag{13}$$

Where w_i is ratio of child node to the parent node.

The structure of tree is parametrised on the basis of the feature selection splitting position of nodes and depth of the tree. The feature which gives low variance or less error is selected, and splitting position is also selected for minimum error criteria [22]. The depth of the tree or the number of leaves is useful for reducing the risk of overfitting. Figure 5.4 shows a flow chart of decision tree regression where every node is split by asking a less than or greater than query. In this chapter, we used the maximum of three leaves per node.

5.3.1.2 Implementation of Random Forest Regression Algorithm

Random forest regression (RFR) is a tree-structured supervised machine learning algorithm used for regression-type problems. The structure of the random forest regression is made by randomly selected dataset cells. The parametrisation technique used for random forest regression is the same as decision tree regression [23, 24] but it also incorporates a sampling technique called Bootstrap Aggregation. In random forest regression, multiple decision trees are constructed in parallel; doing this creates multiple domains with replacement using bootstrapping [25]. Multiple decision trees give multiple outcomes, and when all these outcomes are combined, this technique is called Bagging. In this chapter, the number of trees used for modelling random forest regression is 100.

5.3.1.3   Implementation of Extreme Gradient Boosting Algorithm

Extreme Gradient Boosting (xGBoost) is an ensemble machine learning algorithm [8] used for regression as well as classification problems. It constructs boosted trees in a parallel and efficient manner [26–28]. xGBoost is an advanced form of random forest algorithm, where the tree is constructed using Bootstrapping and Bagging techniques. It uses the Gradient Boosting technique for optimisation.

The goal of this algorithm is to minimise the loss function and construct a more efficient tree than prior model. The formula for loss function of prior model *f(i)* is given in equation (14), and objective function is given in (15).

$$f(i) = \min_y \sum\nolimits_{i=1}^{n} (y_{i\,predicted}, y_i) \tag{14}$$

$$f(obj) = \sum\nolimits_{i=1}^{n} f(i) + \sum\nolimits_{k=1}^{k} f(k) \tag{15}$$

where *f(k)* indicates an independent tree in *f(obj)*, given I is sample number of k_{th} trees. The independent tree *f(k)* is given by:

$$f(k) = \alpha T + \frac{1}{2}\lambda \| w \|^2 \tag{16}$$

where α is learning rate and λ are regularisation parameters, T and *w* are the numbers of leaves and weight of the leaf respectively. In this study, the learning rate is taken as 0.1, the maximum number of leaves is three and trees is 100.

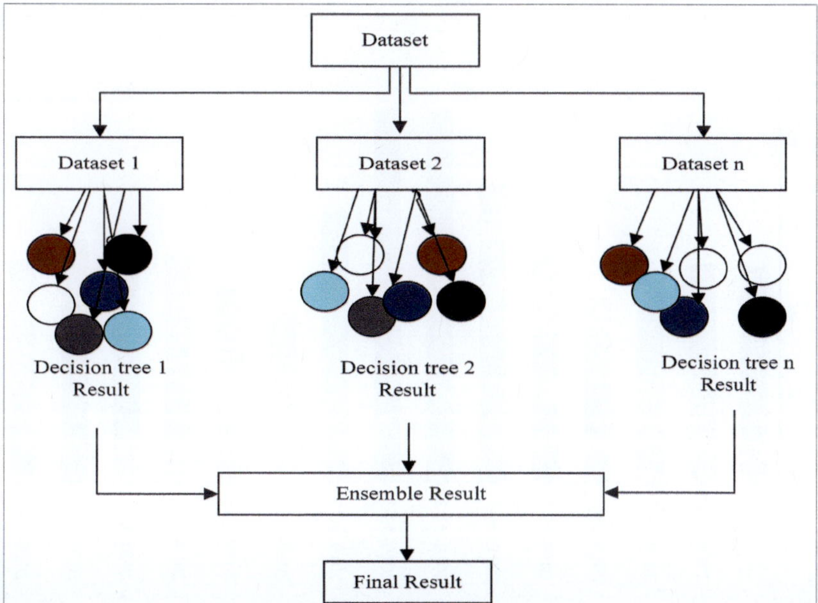

FIGURE 5.5   Random forest regression tree.

5.4   RESULT AND DISCUSSION

5.4.1   DATA DESCRIPTION

Wind speed data with a time resolution of 10 minutes from January 1st, 2019, to March 1st, 2020, has been used in this present work for detailed analysis. The first 70% of the data points served as training samples, while the remainder served as test data. Date and time, ambient temperature, wind direction, rotor RPM and wind speed are all taken into account.

The data features inclusive of maximum wind speed, temperature and rotor RPM are given in Table 5.1, and the average value of wind speed is 5.50 m/s. Monthly and weekly trends of wind speed from January 2019 to March 2020 are shown in Figures 5.6 and 5.7, respectively.

In the months of July, August and September there is a spike in the wind speed dataset.

In Figure 5.8, the bar chart shows that the average wind speed was more than 8 m/s throughout the months of July and August. The maximum power can be generated during these periods.

TABLE 5.1
Description of Dataset

Data features	Count	Mean	Std. mean	25%	50%	Max
Ambient temperature	4409	29.15	4.64	25.61	28.70	42.40
Rotor RPM	4409	9.25	4.81	8.08	9.61	16.07
Wind direction	4409	201.06	85.39	160	182	357
Wind speed	4409	5.60	2.50	3.65	5.30	22.97

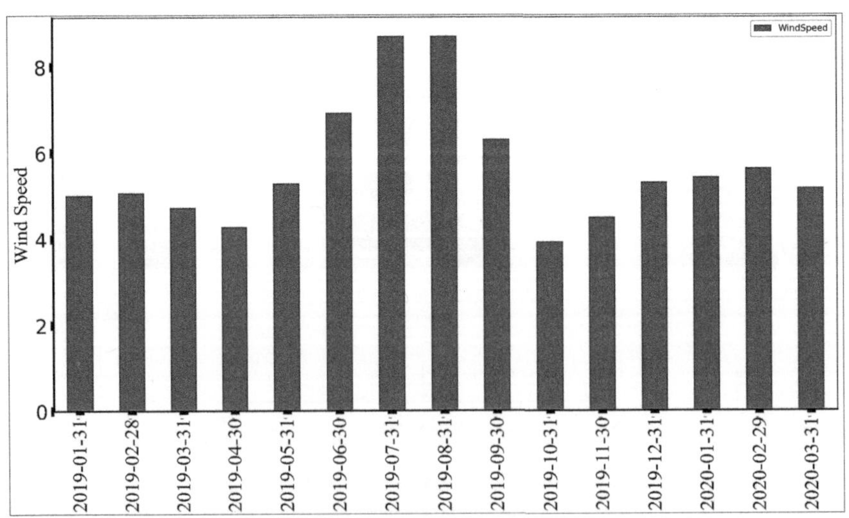

FIGURE 5.6   Monthly wind speed.

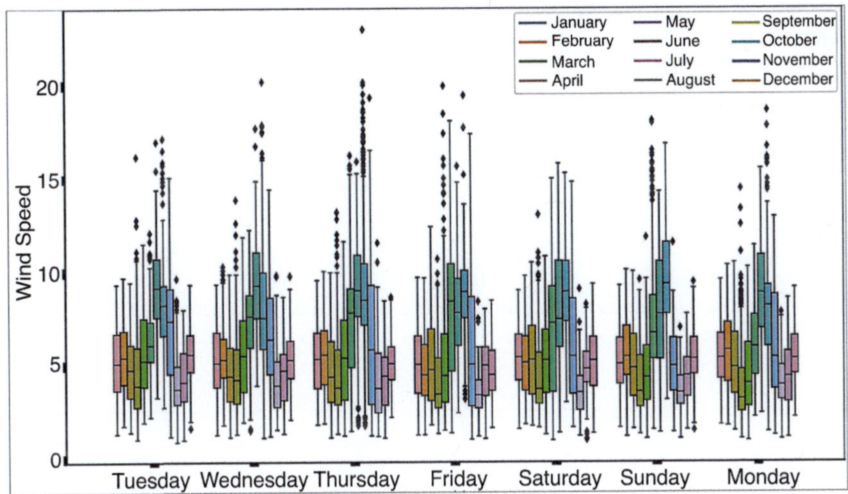

FIGURE 5.7   Weakly trend of wind speed.

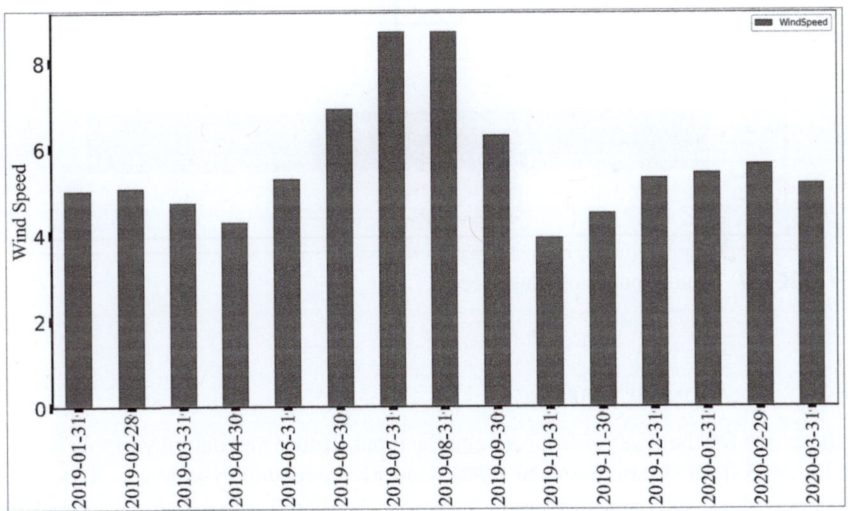

FIGURE 5.8   Monthly wind speed.

5.4.2   DATA PRE-PROCESSING

Data pre-processing entails data cleansing land outlier detection. Dealing with anomalous data samples in the original wind data is outlier detection. The number of null samples for each feature after and before data pre-processing is shown in Table 5.2.

The wind speed of more than 20 m/s is an outlier point, and it can give false results and affect the prediction accuracy. Dropping this value from the dataset will give good prediction accuracy.

TABLE 5.2

Null Data Samples Count

Features	Before	After
Ambient temperature	24407	0
Rotor RPM	56097	0
Wind speed	45946	0
Wind Direction	23629	0
Date–Time	0	0

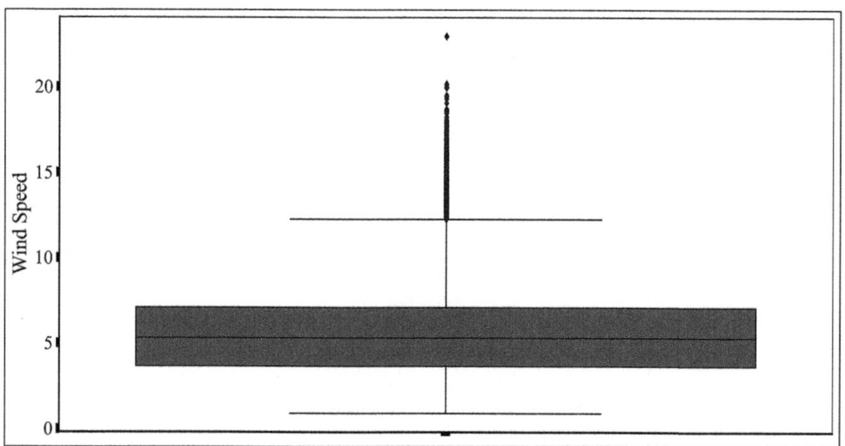

FIGURE 5.9   Outlier points of wind speed.

5.4.3 RELATIONSHIP LEARNING

A heat map for the relationship between the input feature and output variable is given in Figure 5.10. It shows that rotor RPM is positively related to wind speed and wind direction. The ambient temperature is negatively related to wind speed, rotor RPM and wind direction.

5.4.4 PERFORMANCE MATRIC

To evaluate the developed model's accuracy, a number of assessment criteria are used. The Mean Squared Error (MSE), Mean Absolute Percentage Error (MAPE) and Root Mean Square Error (RMSE) [29, 30] are commonly used metrics for determining the aims and outputs. Accuracy is higher when MSE is smaller. Cross Validation Score (CVS) is also used to validate machine learning methods. The list of performance indications used in this chapter is shown in Table 5.3.

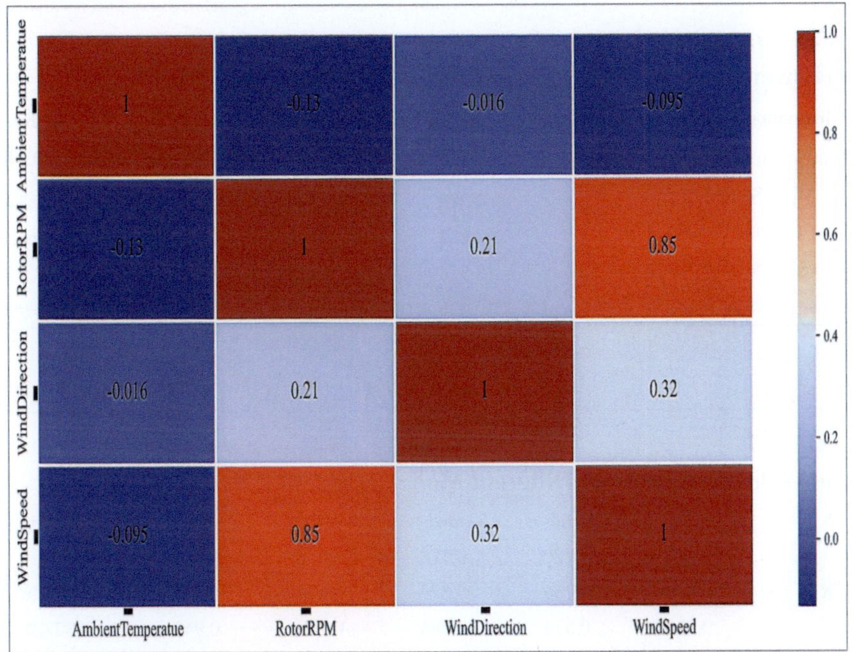

FIGURE 5.10 Relationship between data features.

TABLE 5.3

Equation for Performance Measure

MSE

$$\frac{1}{N}\sum_{i=1}^{n}(y_i - y_{i\,predicted})$$

MAE

$$\frac{1}{N}\sum_{i=1}^{n}\|y_i - y_{i\,predicted}\|$$

MAPE

$$\frac{1}{N}\sum_{i=1}^{n}\|\frac{(y_i - y_{i\,predicted})}{y_i}\|*100$$

RMSE

$$[\frac{1}{N}\sum_{i}^{n}y_i - y_{i\,predicted})^2]^{\frac{1}{2}}$$

R^2

$$1-\frac{MSE(model)}{Average\;value\;of\;actual\;output}$$

CVS

$$\frac{1}{K}\sum_{i=1}^{n}MSE$$

TABLE 5.4

Forecasting Results of Different Model

Performance matrix (Accuracy)	LR	DTR	RFR	xGBoost
Mean Square Error	1.659	0.988	0.565	0.548
Root Mean Square Error	1.288	0.994	0.746	0.740
Mean Absolute Error	0.882	0.566	0.434	0.428
Mean Absolute Percentage Error	18.558	11.815	9.153	8.994
R square error is	0.635	0.839	0.902	0.906

TABLE 5.5

Cross Validation Score of Different Model

CVS	1st	2nd	3rd	4th	5th	Percentage
LR	0.680	0.720	0.718	0.456	0.790	67.3%
DTR	0.834	0.786	0.829	0.674	0.870	79.9%
RFR	0.905	0.874	0.883	0.768	0.940	87.4%
xGBoost	0.914	0.876	0.885	0.785	0.946	88.1%

Where y_i and $y_{ipredicted}$ are the ith actual value and corresponding predicted value. N is the count of samples and K is the division of dataset. In this chapter, K is taken as 5 for *CSV*.

Tables 5.4 and 5.5 exhibit the outcomes of the models used for wind speed/wind power forecasting. It is clear that linear regression performs less well than other alternative methods. Random forest regression and xGBoost perform effectively and produce quality outcomes.

5.5   CONCLUSION

Wind energy exploitation and utilisation can help to mitigate energy shortages and environmental pollution. The most direct way to use wind energy is to convert it to electricity using a wind turbine and then integrate it to the power grid. However, the intermittent nature of wind poses a significant threat to supply stability. To address this issue, appropriate models for generating accurate wind speed/wind power forecasts must be developed. In this chapter, different Machine learning models are used to forecast wind speed.

This study shows that linear regression is fast but gives low accuracy results. The issue with the linear regression model is that it heavily depends on the relation between the input and output features. Random forest regression and Extreme Gradient Boosting are observed as the best among the other illustrated algorithms. These algorithms work well even when the dataset has missing values. The accuracy of random forest regression increases when the number of input features increases.

REFERENCES

1. B. H. Khan, "Nonconventional Energy Sources," second edition, Tata Mc-Graw Hill, pp. 198–200, 2009.
2. Halil Demolli, Ahmet Sakir Dokuz, Alpher Ecemis, and Murat Gokcek, "Wind power forecasting based on daily wind speed data using machine learning algorithms," *Energy Conversion and Management* Vol. 198, p. 111823, 2019.
3. Xiaochen Wang, Peng Guo, and Xiaobin Huang, "A review of wind power forecasting models," *Energy Procedia* Vol. 12, pp. 770–778, 2011.
4. Yun Wang, Runmin Zou, Fang Liu, Lingjun Zhang, and Qianyi Liu, "A review of wind speed and wind power forecasting with deep neural networks," *Applied Energy*, Vol. 304, p. 117766, 2021.
5. Michael Milligan, Kevin Porter, et al., "Wind power myths debunked," *IEEE Power and Energy Magazine* Vol. 7, No. 6, pp. 89–99, 2009.
6. Saurabh S. Soman, Hamidreza Zareipour, Om Malik, and Paras Mandal, "A review of wind power and wind speed forecasting methods with different time horizons," In North American Power Symposium (NAPS). *IEEE*, pp. 1–8, 2010.
7. Jaesung Jung, and Robert P. Broadwater, "Current status and future advances for wind speed and power forecasting." *Renewable and Sustainable Energy Reviews*, 31, pp. 762–777, 2014.
8. Najmeddine Dhieb, Hakim Ghazzai, Hichem Besbes, and Yehia Massoud, "Extreme gradient boosting machine learning algorithm for safe auto insurance operations," *IEEE International Conference on Vehicular Electronics and Safety (ICVES)*, pp. 1–5, September 2019.
9. Minghua Chen, Qunying, Liu, Shuheng Chen, et al., "XGBoostbased algorithm interpretation and application on post-fault transient stability status prediction of power system," *IEEE Access*, Vol. 7, pp. 13149–13158, 2019.
10. N. A. Madlool, M. S. Hossain, et al., "Investigation on wind energy for grid connection in Bangladesh: Case study," *In IOP Publishing Conference Series: Materials Science and Engineering*, Vol. 1127, No. 1, p. 012021, 2021.
11. M. I. Jordan, and T. M. Mitchell, "Machine learning: Trends, perspectives, and prospects," *Science*, Vol. 349, pp. 255–260, 2015.
12. Kathrine Lau Jørgensen, Hamid Reza Shaker, "Wind power forecasting using machine learning: State of the art, trends and challenges," *IEEE 8th International Conference on Smart Energy Grid Engineering*, pp. 44–50, 2020.
13. H. Roopa and T. Asha, "A linear model based on principal component analysis for disease prediction," *IEEE Access*, Vol. 7, pp. 105314–105318, 2019.
14. Subham Shaw, and M. Prakash, "Solar radiation forecasting using support vector regression." *IEEE International Conference on Advances in Computing and Communication Engineering (ICACCE)*, pp. 1–4, 2019.
15. Dastan Maulud, and Adnan Mohsin Abdulazeez, "A review on linear regression comprehensive in machine learning," *Journal of Applied Science and Technology Trends*, Vol. 1, No. 4, pp. 140–147, 2020.
16. Yoshihisa Fujita, Soichiro Ikuno, Taku Itoh, and Hiroaki Nakamura, "Modified improved interpolating moving least squares method for meshless approaches," *IEEE Transactions on Magnetics*, Vol. 55, pp. 1–4, 2019.
17. Manuel Herrera, Luis Torgo, Joaquin Izquierdo, and Rafael Perez-Garcia, "Predictive models for forecasting hourly urban water demand," *Journal of Hydrology*, Vol. 387, No. 1–2, pp. 141–150, 2010.
18. Qi Wang, Yue Ma, Kun Zhao, and Yingjie Tian, "A comprehensive survey of loss functions in machine learning," *Annals of Data Science*, Vol. 9, No. 2, pp. 187–212, 2022.
19. Ethem Alpaydin, "Introduction to Machine Learning," third edition, **PHI Learning Pvt. Ltd.**, p. 640, 2014

20. Fetty Fitriyanti Lubis, Yusep Rosmansyah, and Suhono Harso Supangkat, "Gradient descent and normal equations on cost function minimization for online predictive using linear regression with multiple variables," *IEE 2014 International Conference on ICT For Smart Society (ICISS)*, pp. 202–205, 2014.

21. Leon Bottou, "Large-scale machine learning with stochastic gradient descent," *Proceedings of COMPSTAT, Physica-Verlag HD*, pp. 177–186, 2010.

22. Philip A. Chou, "Optimal partitioning for classification and regression trees," *IEEE Transactions on Pattern Analysis & Machine Intelligence*, Vol. 13, No. 04, pp. 340–354, 1991.

23. Aleena Swetapadma, and Anamika Yadav, "A novel decision tree regression-based fault distance estimation scheme for transmission lines," *IEEE Transactions on Power Delivery*, Vol. 32, No. 1, pp. 234–245, 2017.

24 Wei-Yin Loh, "Classification and regression trees," *Wiley Interdisciplinary Reviews: Data Mining and Knowledge Discovery*, Vol.1, No.1, pp. 14–23, 2011.

25. Jan Peters, Bernard De Baets, Niko E.C. Verhoest, et al., "Random forests as a tool for ecohydrological distribution modelling," *Ecological Modelling*, Vol. 207, No. 2–4, pp. 304–318, 2007.

26. Jianming Hu, Jiani Heng, Jiemei Wen, and Weigang Zhao, "Deterministic and probabilistic wind speed forecasting with de-noising-reconstruction strategy and quantile regression-based algorithm," *Renew Energy*, Vol. 162, pp. 1208–26, 2020.

27. H.Z. Wang, G.B. Wang, G.Q. Li, J.C. Peng and Y.T. Liu, "Deep belief network based deterministic and probabilistic wind speed forecasting approach," *Applied Energy*, Vol. 182, pp. 80–93, 2016.

28. Zhenhai Guo, YaoDong, Jianzhou Wang, and Haiyan Lu, "The forecasting procedure for long-term wind speed in the Zhangye area," *Mathematical Problems in Engineering*, Vol. 2010, p. 684742, 2010.

29. Subham Shaw, and M. Prakash, "Forecasting solar potential using support vector regression." *IEEE Devices for Integrated Circuit (DevIC)*, pp. 5–8, 2019.

30. P. Marimuthu, P. Chinnamuthu, and R. Jeyapaul, "Analysis of parameter selection for solar radiation prediction and global solar radiation prediction model using polynomial regression", *International Journal of Mathematics in Operational Research*, Vol. 16, No. 4, pp. 469–479, 2020.

6 IOT Communication Technologies

Dr Shaik Kareemulla and Dr Syed Khasim

6.1 THE INTERNET OF THINGS (IOT): WHAT IS IT?

The "internet of things" (IoT) is a network of devices that can exchange data with one another and potentially with remote cloud-based servers. Items with embedded software and sensors, consumer goods, and mechanical and digital gear are a few examples of the kinds of things that can be a part of the IoT. Businesses of various sizes and in a variety of industries are utilizing the Internet of Things to increase output, customer happiness, and return on investment. Data can be moved over a network via the Internet of Things [1] without requiring human or human-computer contact.

In the context of the IoT, "things" include the person wearing an implanted cardiac monitor, the farm animal with a biochip transponder, the car with integrated sensors alerting the driver to low tire pressure, and any other natural or artificial object that can obtain an IP address and transmit data via a network.

6.1.1 HOW DOES THE INTERNET OF THINGS WORK?

In an IoT world, web-enabled smart devices use embedded systems, which are made up of CPUs, sensors, and communication gear [2], to get information from their surroundings, send it to other devices, and act on it.

When connected to an IoT gateway, sensors on IoT devices can transfer the data they've collected to a central location. Data can be delivered to an edge device for local analysis before being shared [1]. Bandwidth needs are reduced when data is analyzed locally rather than being uploaded to the cloud.

A number of these gadgets can talk to each other and react to the information they send and receive. The devices can be used by people, but most of the work is done electronically to set them up, teach them, or look at their records, among other things. IoT applications greatly affect the networking, gathering, and communication protocols [2] that these web-enabled devices use. The application of AI and ML within the IoT can also facilitate the simplification and adaptation of data collection procedures in Figure 6.1 [3].

Data from sensors built into IoT devices is gathered by an IoT system and sent to an IoT gateway before being sent to an application or back-end system for analysis.

DOI: 10.1201/9781003452393-6

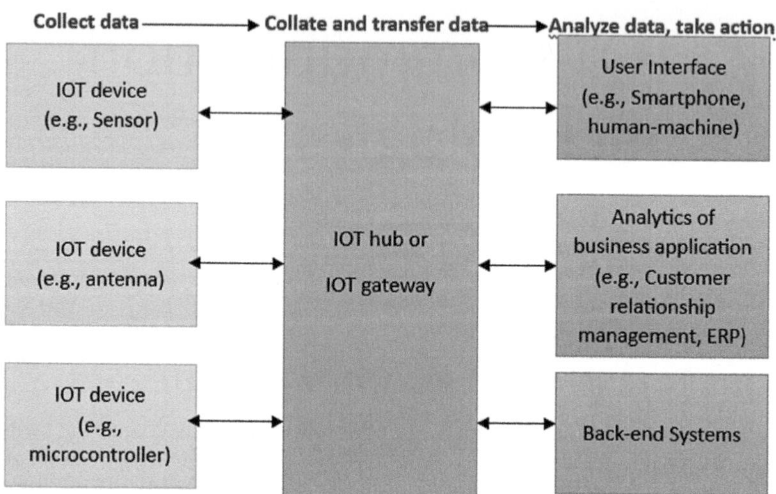

FIGURE 6.1 Example of an IOT system.

6.1.2 WHY IS IoT IMPORTANT?

The Internet of Things makes daily life and productivity more efficient. For example, consumers can benefit from IoT-enabled automobiles, smart watches, and home thermostats. A person's car could notify the garage to open when they pull up, the thermostat could drop the temperature to a comfortable level, and the lights could dim or change colour to welcome them home.

IoT is important not just for home automation but also for commercial enterprises, thanks to the smart gadgets it delivers. It gives businesses an immediate picture of how their systems function, from machine efficiency to supply chain and logistics management.

In a world where everything is connected, machines can do our dirty work for us. Businesses may save money on labour, decrease waste, and provide better service all by using automation. With the help of IoT, production and shipping costs can be reduced, and businesses gain insights into their customers' buying habits.

6.1.3 WHEN IT COMES TO BUSINESSES, WHAT ARE THE ADVANTAGES OF IoT?

The IoT has several uses for businesses. Some benefits are exclusively applicable to one industry, while others can be used in several. These are some of the most frequently cited advantages for companies:

- Keeps an eye on things from a high level.
- Raises satisfaction levels for the company's clientele.
- It helps you avoid wasting both money and time.
- Boosts efficiency and output from workers.
- Offers flexibility and integration for enterprises.
- It helps companies make smarter choices.
- Increases income.

IoT provides organizations with the impetus and resources to re-evaluate and refine their operations and methods.

Sensors and other Internet of Things devices are most used in the manufacturing, transportation, and utility industries, but there are also growing markets for these technologies in the areas of home automation and agricultural.

For farmers, the Internet of Things may make farming easier. By collecting information on weather, humidity, temperature, and soil composition, IoT and sensors can automate farming.

Infrastructure operations can be monitored with the use of IoT as well. Sensors, for instance, can keep an eye out for anything that can undermine the integrity of a building, bridge, or other infrastructure. Advantages include faster response times to incidents, lower operational expenses, and a higher quality of service [4]. Using the Internet of Things, a home automation company can keep tabs on and adjust the home's electrical and mechanical components. In a larger sense, smart cities can aid residents in cutting down on garbage and utility bills. The IoT affects every sector of the economy.

6.1.4 THE INTERNET OF THINGS: PROS AND CONS

The following are a few benefits of the IoT:

- Allows users to get their hands on data whenever and wherever they want it.
- Facilitates enhanced interaction between electrical gadgets.
- Allows data packets to be sent and received through a network, which can be a time- and cost-saver.
- It helps consumers and producers by compiling massive volumes of data from a variety of devices.
- Reduces the quantity of information that must be uploaded to the cloud by performing analysis locally.
- Reduces the demand for human labour while simultaneously enhancing a company's service quality through automation.
- Allows for more consistent and efficient healthcare delivery.

Amazon is well-represented in the cloud computing industry by means of Amazon Web Services (AWS) IoT. This framework was created to provide rapid and secure connectivity between intelligent devices and the Amazon cloud, enabling them to exchange data.

Arm microcontrollers can be used to build Internet of Things applications via the open-source Arm Mbed IoT framework. By merging Mbed services and technologies, this Internet of Things platform provides IoT devices with a scalable, networked, and secure environment.

The Microsoft Azure IoT Suite platform is a collection of services that let customers interact with their IoT devices, process and receive data, and view data processing activities that are beneficial to businesses, like aggregation, transformation, and multidimensional analysis.

The Calvin open-source Internet of Things platform from Ericsson is designed for developing and administering distributed applications that enable device-to-device communication.

Calvin provides a runtime environment and programming framework for creating and running IoT applications for businesses and consumers.

The Internet of Things offers a wide range of real-world applications, from consumer and commercial to institutional and industrial.

Among the sectors that stand to gain from the IoT are automotive, communications, and energy. With "smart homes" that include appliances and thermostats, customers may use their computers or mobile devices to remotely monitor and control their home's electronics, lighting, and temperature from anywhere. Wearable technology that has sensors and software built in can collect and analyze user data and share that data with other technologies to enhance the user experience. In public safety, wearables are used to monitor the health of first responders in dangerous situations, such as during construction or fires, or to improve their response times during emergencies. By analyzing the data generated by IoT devices, healthcare providers may keep a closer eye on their patients.

Inventory management of drugs and medical equipment is only one example of how frequently IoT technologies are used in hospitals. One way in which smart buildings save money is by employing occupancy sensors to adjust the heating and cooling accordingly. Sensors can determine when a room is full and activate the air conditioner or when the office is empty and reduce the temperature accordingly. Connected sensors in smart farming systems built on the IoT can keep an eye on the field's illumination, temperature, humidity, and soil moisture. Automating irrigation systems is another useful application of IoT. Two examples of IoT sensors and deployments that can improve sanitation, lessen traffic, monitor and manage environmental concerns, and inform citizens are smart meters and lighting.

6.1.5 Problems in Protecting User Data and Ensuring Device Integrity in the IoT

In 2016, one of the most publicized Internet of Things assaults ever occurred. The domain name server provider Dyn was compromised by the Mirai botnet, which led to significant downtime. Insecure Internet of Things (IoT) devices were exploited by attackers to obtain access to the network. Even now, Mirai is being refined to become one of the most devastating DDoS attacks ever witnessed.

Owing to the interconnectedness of IoT devices, an attacker can corrupt all the data and destroy the system with just one security flaw [4, 5]. When manufacturers don't provide regular updates for their products, hackers can exploit those products. Hackers are interested in personal data that linked devices frequently seek, such as names, ages, addresses, phone numbers, and even social network profiles.

However, hackers aren't the only cause for fear when it comes to the IoT. Companies involved in the production and distribution of consumer IoT devices might, for instance, harvest and resell users' personally identifiable information.

6.2 CONNECTIVITY IN THE IOT

Connected devices over the internet can share information and carry out activities autonomously thanks to the IoT. Electronics, software, a network, and sensors all play a role in facilitating communication with these gadgets. To ensure the success of an IoT product or project, it is crucial that connected devices be able to communicate with one another.

6.2.1 SOME EXAMPLES OF IoT COMMUNICATION TYPES ARE LISTED HERE

6.2.1.1 H2M (Human to Machine)

Here, a person provides directions to an IoT gadget in the form of data (spoken/written/visual/etc.). Machines that are part of the IoT, such as sensors and actuators, can take human input, process it, and give feedback in the form of text or a graphical user interface. This is quite helpful because machines like this help people with everything. It's a combination of software and technology that allows a person to work in tandem with a machine to get something done, which is shown in Figure 6.2.

The advantages of this H2M include its straightforward interface and the ease with which it can be accessed by the user. Faster reaction times to problems mean less downtime. It has several customizable options and features.

- Instances include face recognition systems.
- Sign-in with biometrics.
- Recognizing a person's speech or voice.

FIGURE 6.2 H2M communication.

6.2.1.2 Machine-to-Machine (M2M)

The process of sending messages or data between two or more computers or devices is referred to by this term. It is the exchange of physical objects with one another that does not need a middleman. M2M communication is sometimes referred to as machine-type communication by the 3GPP. Here, machines talk to one another through the exchange of preprogrammed data. For machines to talk to one another, they need instructions. Here, people talk to each other without actually talking to each other. Both wired and wireless connections can be used to link the machines. An M2M connection is a direct link established between two network nodes to facilitate data transmission across public networking protocols such as Ethernet or cellular. IoT builds upon the foundations of M2M by connecting disparate electronic devices into vast "cloud" networks that exchange data and instructions via remote servers, shown in Figure 6.3.

M2M's benefits include its ease of administration and ability to operate over cellular networks. It can be applied in multiple contexts and facilitates hands-free communication between intelligent objects. The M2M contact facility is the best place to address security and privacy concerns related to IoT networks [6, 7]. Large-scale data collection, processing, and protection are feasible. Nevertheless, a disadvantage of cloud computing in M2M is that it restricts flexibility and creativity. The safety and ownership of the data are of paramount importance. Interoperability between cloud and M2M IoT solutions presents a significant hurdle. Connectivity between M2M devices requires constant access to the internet.

Clothes washed or dried in a smart washing machine, for instance, will trigger an alert on the owner's smartphone. Smart meters can monitor and report on energy consumption in homes and businesses.

6.2.1.3 Machine to Human (M2H)

Here, the machine communicates with people. The machine initiates communication (text, images, audio, or visual signals) in response to or independently of human presence. This mode of interaction is typically employed in settings where machines assist people in doing routine tasks. It's a method of collaboration in which people use machines and smart technologies to get things done, as shown in Figure 6.4.

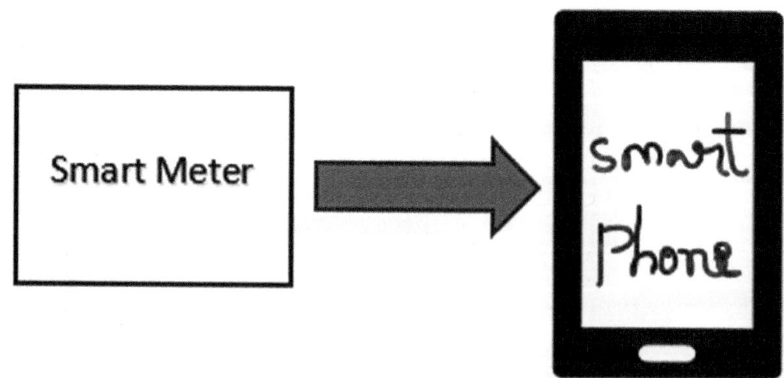

FIGURE 6.3 M2M communication.

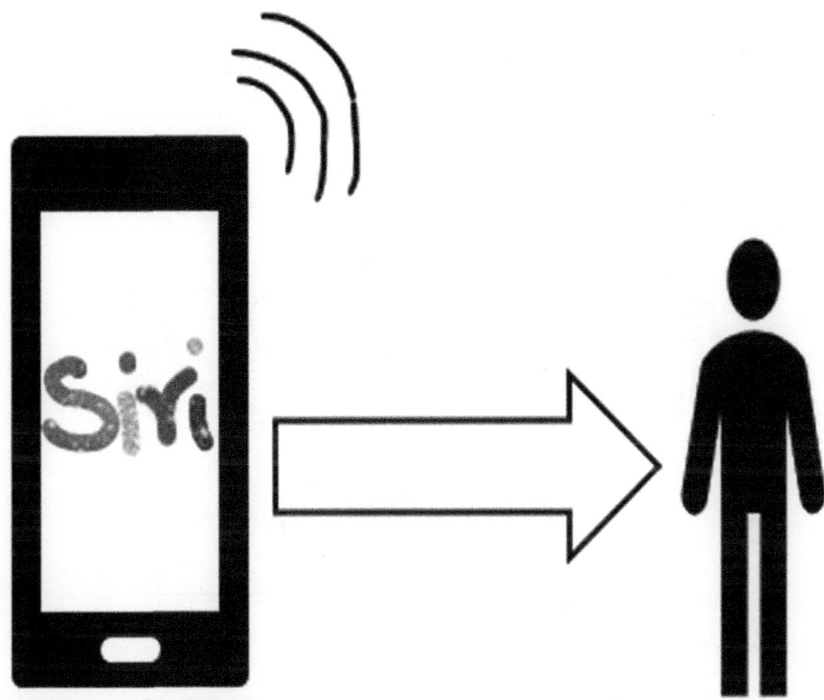

FIGURE 6.4 M2H communication.

6.2.1.3.1 Examples

Traffic flow lights, fitness bands, fire alarms, and health monitoring equipment.

6.2.1.4 Human to Human (H2H)

Humans use a wide variety of methods, including verbalization, graphical representation, facial expression, and bodily language, to convey meaning to one another. Without human intervention to quickly address and manage issues, obstacles, and scenarios, it will be impossible to realize the anticipated benefits of M2M applications.

Human-to-human (H2H) communication refers to the transfer of data or messages between individuals. This can be accomplished in several ways, including through spoken, nonverbal, and written exchanges shown in Figure 6.5.

The Internet of Things devices communicate with one other via a multitude of protocols. The shared data amongst IoT-connected devices is safeguarded by these protocols, which are communication techniques via technology such as Wi-Fi, Bluetooth, ZigBee, and so on.

> **Protocols for Interoperability in the Internet of Things:** In this part, we will go into detail about the various methods of communication. Figure 6.6 depicts two common classifications for IoT communication protocols: (1) LPWAN and (2) short-range network.

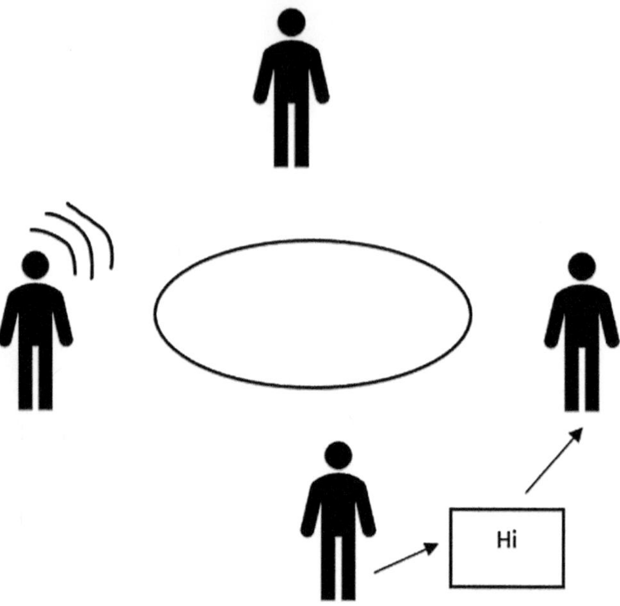

FIGURE 6.5   H2H communication.

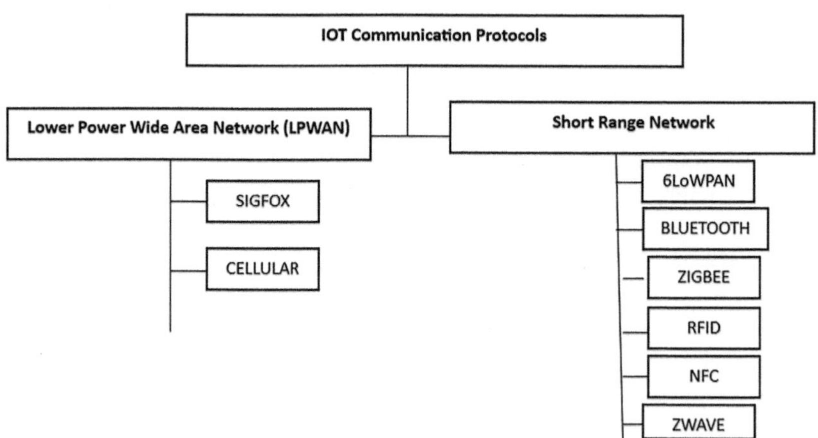

FIGURE 6.6   LPWAN.

Sigfox I.I.A: SigFox is a low-energy technology [8,9] that facilitates M2M and wireless sensor network communication. It's possible to send small data packages up to 50 kilometers. SigFox employs ultra-narrow band (UNB) technology. This technology has an extremely low battery capacity; therefore, data transfer rates are limited in the range 10 and 1,000 bits per second. Smart meters, environmental sensors, agribusiness, safety devices, smart lighting, and patient monitoring all make use of near-field communication (NFC) technology.

Tissue Type II.I.B: IoT applications that need to function across greater distances and make use of large amounts of data should take advantage of cellular technology. Cellular connection technologies such as GSM, 3G, and 4G may provide a stable and fast internet connection. It does, however, demand a lot of energy to operate. Therefore, it cannot be utilized for exchanges over regional networks or even machine-to-machine networks. Cellular communication protocol is widely utilized, particularly in situations where mobile devices are the focus. Multiple types of technology influence the structure of cells [8, 10, 11].

6.3 LIMITED-DISTANCE NETWORK

6.3.1 6LoWPAN

6LoWPAN is the first and current standard for IoT communication protocols since it is an IP-based standard internetworking protocol. It can talk to any other IP network without the help of any adapters, gateways, or proxies. This standard was created by the Internet Engineering Task Force (IETF) to make it easier to use IPv6 over IEEE802.15.4-based low-power wireless networks. The maximum number of IP addresses it can handle is 2128, which is plenty. The reason for this is to accommodate addresses of varied lengths. It uses very little power and costs very little to operate. Many network architectures, such as mesh and star, are supported by 6LoWPAN. 6LoWPAN includes an adaptation layer between the MAC and network layers to handle interoperability between IEEE 802.15.4 and IPv6. Figure 6.2 shows that compared to 6LoWPAN, ZigBee is the superior option. The physical layer protocol used by both systems is IEEE 802.15.4 [12–16].

6.3.2 ZIGBEE

To put into practice the IEEE802.15.4 standard for low-power wireless networks, the ZigBee Alliance created the ZigBee protocol. ZigBee was designed to standardize the usage of small, low-power digital radios capable of long-range data transmission, enabling the cheapest possible implementation of personal area networks. Also, it will be utilized in applications that can't handle a lot of data, need a lot of battery life, and need to be able to trust their network. ZigBee additionally supports several network topologies [8, 12], including mesh, star, and tree topologies.

6.3.3 BLE

Bluetooth Low Energy, or BLE for short, is a crucial protocol for IoT use. It has been optimized for low-latency, low-bandwidth Internet of Things use. BLE traditional Bluetooth has many benefits, including low power consumption, quick setup, and the capability for an infinitely large number of nodes in a star network topology [8, 12].

6.3.4 RFID Gen II

RFID standards have been developed by a wide variety of organizations, including the International Organization for Standardization (ISO), the International Electrotechnical Commission (IEC), ASTM International [17, 18], the DASH7 Alliance, and EPC-Global. In radio frequency identification systems, the RF tags utilized are small radio frequency transponders. Since this tag has been electronically programmed with the necessary information, it may be read from a considerable distance. RFID tag systems can use either an active reader tag system or a passive reader tag system for reading the tags. Passive tags can function at lower frequencies and are less expensive to manufacture than their active, battery-powered counterparts. There is a delay in getting the measurement and diagnostic data using RFID because the data is static and must be entered into the tag. Smart retail, healthcare, national security, and agriculture are just a few examples of IoT applications that make use of RFID technology. RFID enables a network to have a decentralized, peer-to-peer structure [9, 19, 20].

6.3.5 NFC II

Data can be transferred between two devices utilizing near-field communication (NFC) when they are within a few inches of one another or can be touched. Similar technology underpins both RFID and NFC. Identification plays a secondary role as a high-tech form of two-way communication. Although NFC tags can retain information, they have a limited storage capacity. This tag can be written to by the device at a later time, unlike RFID tags, which are read-only and cannot be updated. Reader/ writer, active card emulation, and passive peer-to-peer are the three basic uses for near-field communication. NFC technology has found widespread use in contactless payment systems, industrial applications, and mobile phones. Similarly, NFC makes it easier to set up and manage IoT devices in a variety of environments, such as the home, the office, and the factory. The NFC peer-to-peer technique of device pairing is convenient [9, 12, 19–23].

6.3.6 Z-Wave

Zensys' low-power MAC protocol, which makes use of wireless home automation to link 30–50 nodes, has been utilized for Z-Wave IoT communication, especially in the smart home and small business domains [24, 25]. This technology allows for point-to-point communications up to 30 meters in range, with data packet speeds of up to 100 kbps. Because of this, IoT-based energy, lighting, and healthcare management systems can greatly benefit from the usage of condensed communications. There is always a master and a slave in a Z-Wave network. Slave nodes are inexpensive but passive devices that can only receive data. Only other networked devices may send it commands, and it must obey those commands in order to function. Z-Wave networks facilitate the mesh topology [26–28].

6.3.6.1 IoT Communication Protocols: A Comparative Study of the Third Generation

This section's goal is to give researchers a roadmap for choosing the most effective communication protocol by contrasting the available options. The many means of communication are compared using a wide variety of parameters. Standards, networks, topologies, voltages, ranges, cryptography, spreading, modulation types, coexistence mechanisms, security, power consumption, and more are all taken into account when covering IoT IP.

6.4 IOT IP COVERAGE [29]

There is a high level of security across all nine approaches because of built-in mechanisms for user authentication and data encryption. All the wireless networking protocols (6LoWPAN, ZigBee, BLE, NFC, and Z-Wave) that operate in counter mode employ the Advanced Encryption Standard (AES) block cipher[30, 31, 23]. But RC4 is what you will find in cellular and RFID systems. However, several significant problems were discovered. In contrast, AES is substantially more secure than RC4. When compared to AES, RC4's speed is far more impressive.

Low-power wireless networking technologies include 6LoWPAN, ZigBee, Bluetooth Low Energy, Z-Wave, and Near-Field Communication. This means that it can function on less energy. However, there are additional considerations, such as the energy consumption of portable devices.

For 6LoWPAN, ZigBee, BLE, NFC, SigFox, and Z-Wave, it is less than or equal to 1 Mbps. In contrast, RFID has a staggering data transfer rate of 4 Mbps. The coverage areas for SigFox and cellular networks are larger than a few kilometres. On the other hand, the ranges of technologies like 6LoWPAN, ZigBee, BLE, NFC, Z-Wave, and RFID [32–34] are typically between a few hundred metres and a few kilometres at most.

When comparing IoT communication technologies, 6LoWPAN [10, 15] stands out as the leader because it is IP-based WSN. IPv6's vast address space for data and information collection makes it simple to set up a large number of smart devices over the internet. This is feasible due to several features, including low bandwidth requirements, scalability, low cost, low power consumption, and the availability of different topologies like star and mesh [35, 36].

6.5 SUMMARY

The IoT network incorporates a variety of wireless technologies, each with its own set of requirements and benefits. However, it may be difficult to identify the best alternative. The question that needs answering is, "Which platform is best for my particular application?" The purpose of this study is to accomplish this by comparing and contrasting the most popular IoT communication protocols. Many different metrics are used to assess the efficacy of various communication tools. Some examples of such elements are network topology, power, range, cryptography, spreading, modulation type, mechanism coexistence, and power consumption. Building on existing work, future studies will investigate IoT apps and IoT security methods for real-time detection of IoT assaults, including unique IoT risks, and user notification.

REFERENCES

1. L. Tan and N. Wang, "Future internet: The internet of things", Advanced Computer Theory and Engineering (ICACTE) 2010 3rd International Conference, pp. 5–376, 2010.
2. I. Strategy and P. Unit, ITU Internet Reports 2005: The internet of things. Geneva: International Telecommunication Union (ITU), 2005.
3. S. J. Bigelow, Ultimate IoT implementation guide for businesses. London: IoT Agenda, 2021.
4. X. Li, Z. Xuan and L. Wen, "Research on the architecture of trusted security system based on the internet of things", Intelligent Computation Technology and Automation (ICICTA) 2011 International Conference, pp. 1172–1175, 2011.
5. M. Amin, M. Reaz, J. Jalil and L. Rahman, Digital modulator and demodulator IC for RFID tag employing DSSS and Barker code. Journal of Applied Research and Technology, vol. 10, no. 6, pp. 819–825, 2012.
6. G. C. Garc'ia, I.L. Ruiz and M. A. Gomez-Nieto, State of the art trends and future of bluetooth low energy near field communication and visible light communication in the development of smart cities. Sensors, vol. 16, no. 11, 2016.
7. V. Alarcon-Aquino, M. Dominguez-Jimenez and C. Ohms, Design and implementation of a security layer for RFID systems. Journal of Applied Research and Technology, vol. 6, no. 2, pp. 69–82, 2008.
8. F. Samie, L. Bauer and J. Henkel, "IoT technologies for embedded computing: A survey", Hardware/Software Codesign and System Synthesis (CODES+ ISSS) 2016 International Conference, pp. 1–10, 2016.
9. S. Al-Sarawi, M. Anbar, K. Alieyan and M. Alzubaidi, "Internet of Things (IoT) communication protocols", 8th International conference on information technology (ICIT), pp. 685–690, IEEE, 2017, May.
10. L. Alliance, A technical overview of LoRa and LoRaWAN. White Paper, November 2015.
11. M. S. Hossen, A. F. M. Kabir, R. H. Khan and A. Azfar, Interconnection between 802.15. 4 devices and IPv6: implications and existing approaches. arXiv preprint arXiv:1002.1146, 2010.
12. J. Granjal, E. Monteiro, and J. Sa Silva, "Security for the internet of things: A survey of existing protocols and open research issues," in IEEE Communications Surveys Tutorials, vol. 17, no. 3, 2015, pp. 1294–1312.
13. A. Mart'inez Aragues, I. Del, P. Valle, P. Munoz, J. Escayola and J. D. Trigo, Trends in entertainment home automation and e-health: Toward cross-domain integration. IEEE Communications Magazine, vol. 50, no. 6, 2012.
14. A. Al-Fuqaha, M. Guizani, M. Mohammadi, M. Aledhari and M. Ayyash, Internet of things: A survey on enabling technologies protocols and applications. IEEE Communications Surveys & Tutorials, vol. 17, no. 4, pp. 2347–2376, 2015.
15. A. B. Ab Rahman and R. J. Azamuddin, Comparison of internet of things (IoT) data link protocols (pp. 1–21). Technical report, Technical report, Washington University in St Louis, 2015.
16. C.-W. Lu, S.-C Li and Q. Wu, "Interconnecting ZigBee and 6LoWPAN wireless sensor networks for smart grid applications", Sensing Technology (ICST) 2011 Fifth International Conference, pp. 267–272, 2011.
17. A. Ahmad, Wireless and mobile data networks. Hoboken, NJ: John Wiley & Sons, 2005.
18. C. Gomez, J. Oller and J. Paradells, Overview and evaluation of bluetooth low energy: An emerging low-power wireless technology. Sensors, vol. 12, no. 9, pp. 11734–11753, 2012.
19. R. Porkodi and V. Bhuvaneswari, "The Internet of Things (IoT) applications and communication enabling technology standards: An overview", Intelligent Computing Application (ICICA) 2014 International Conference, pp. 324–329, 2014.

20. M. Paavola, Wireless technologies in process automation-review and an application example. Oulu: Control Engineering Laboratory University of Oulu, 2007.
21. A. Le, J. Loo, A. Lasebae, M. Aiash and Y. Luo, A study on QoS security threats and countermeasures using intrusion detection system approach. International Journal of Communication Systems, vol. 25, no. 9, pp. 1189–1212, 2012.
22. M. Zemede, Explosion of the Internet of Things: What does it mean for wireless devices. New York: Keysight Technologies, 2015.
23. L. Alliance, A technical overview of LoRa and LoRaWAN. White Paper, November 20, 2015.
24. J. Hughes, J. Yan and K. Soga, Development of wireless sensor network using bluetooth low energy (BLE) for construction noise monitoring. International Journal on Smart Sensing and Intelligent Systems, vol. 8, no. 2, pp. 1379–1405, 2015.
25. R. Sanchez-Iborra and M.-D. Cano, State of the art in LP-Wan solutions for industrial IoT services. Sensors, vol. 16, no. 5, 2016.
26. C. Goursaud and J. Gorce, Dedicated networks for IoT: PHY/MAC state of the art and challenges. London: EAI Endorsed Transactions on Internet of Things, 2015.
27. C. Gomez and J. Paradells, Wireless home automation networks: A survey of architectures and technologies. IEEE Communications Magazine, vol. 48, no. 6, 2010.
28. A. D. Rathnayaka, V. M. Potdar and S. J. Kuruppu, "Evaluation of wireless home automation technologies", Digital Ecosystems and Technologies Conference (DEST) 2011 Proceedings of the 5th IEEE International Conference, pp. 76–81, 2011.
29. P. Friess, Internet of things: Converging technologies for smart environments and integrated ecosystems. London: River Publishers, 2013.
30. P. Lopez, D. Fernandez, A. J. Jara and A. F. Skarmeta, "Survey of internet of things technologies for clinical environments", Advanced Information Networking and Applications Workshops (WAINA) 2013 27th International Conference, pp. 1349–1354, 2013.
31. R. Tabish, A. B. Mnaouer, F. Touati and A. M. Ghaleb, "A comparative analysis of BLE and 6LoWPAN for U-HealthCare applications", GCC Conference and Exhibition (GCC) 2013 7th IEEE, pp. 286–291, 2013.
32. M. Kuzlu, M. Pipattanasomporn and S. Rahman, "Review of communication technologies for smart homes/building applications", Innovative Smart Grid Technologies-Asia (ISGT ASIA) 2015 IEEE, pp. 1–6, 2015.
33. S. S. I. Samuel, "A review of connectivity challenges in IoT-smart home", Big Data and Smart City (ICBDSC) 2016 3rd MEC International Conference, pp. 1–4, 2016.
34. U. Raza, P. Kulkarni and M. Sooriyabandara, Low power wide area networks: An overview. IEEE Communications Surveys & Tutorials, vol. 19, no. 2, pp. 855–873, 2017.
35. T. M. Shah. Shreya, Security of NFC Data. International Journal of Advanced Research in Computer Science and Software Engineering, vol. 6, n.d.
36. V. Alarcon-Aquino, M. Dominguez-Jimenez and C. Ohms, Desing and implementation of a security layer for RFID systems. Journal of Applied Research and Technology, vol. 6, No. 2, pp. 69–82, 2008.

7 Machine Learning Models for Intelligent Hazard Management

Priyadharshini Ravi and Dr. Senthil Janarthanan

7.1 INTRODUCTION

In today's world, where everything is constantly changing and connected, managing hazards has become a crucial aspect of our lives. It is essential to predict, evaluate, and respond to various hazards in order to protect people, assets, and the environment. Machine learning models are algorithms that can identify patterns or make predictions on unseen datasets. They can be used to forecast extreme events, develop hazard maps, detect events in real-time, provide situational awareness and more decision support. By using these models, we can reduce the risk associated with disasters and make our world a safer place (Cutter *et al.*, 2008). The central objective of hazard management is to minimize the impact of hazards and optimize responses to emerging threats. It is a crucial challenge that transcends domains and boundaries. In the environmental sphere, it involves predicting and alleviating natural disasters such as floods, wildfires, and earthquakes (Paton *et al.*, 2008). Machine learning provides a valuable, data-driven, forward-thinking, and flexible approach to effectively dealing with hazards, making it an essential tool in the quest for intelligent hazard management (Vayà *et al.*, 2013). Machine language is a powerful tool that can be used to enhance hazard management in several ways. Some of those ways are improving the precision and timeliness of hazard predictions, recognizing and evaluating susceptibilities to potential risk, efficiently distributing resources for hazard prevention, readiness, response, and recovery, and enhancing collaboration and information sharing among involved parties (Pradhan *et al.*, 2019). The primary objective of this chapter is to elucidate the intricate nexus between machine learning and the field of hazard management. We will embark on a comprehensive exploration of the foundational principles of machine learning, spanning the spectrum from supervised to unsupervised learning techniques to the essential processes of data acquisition and preprocessing. Our intellectual odyssey will progress with a thorough and comprehensive analysis of predictive models and the indispensable facet of anomaly detection, emphasizing their wide-ranging utility across diverse hazard domains. Furthermore, we will conduct an in-depth examination of the functions played by geospatial analysis and the critical support within the context of machine learning models. Additionally, we will underscore the significance of

DOI: 10.1201/9781003452393-7

human expertise in providing guidance and direction to these machine learning systems. Our discourse will feature illustrative case studies hailing from diverse domains, meticulously accentuating the achievements, obstacles, and knowledge gleaned from these real-world applications. To conclude, we will embark upon a forward-looking contemplation, delving into emerging trends and the dynamically shifting challenges that lie ahead within the domain of machine learning for the facilitation of intelligent hazard management. Through interdisciplinary collaboration, ethical considerations, and innovative research, we envision a future where machine learning is an indispensable tool in safeguarding our world from diverse hazards (Gaber *et al.*, 2005).

7.2 KNOWN HAZARD MANAGEMENT

Hazard management is a pivotal component in safeguarding the safety and welfare of individuals, communities, and the environment. This section delves into the fundamental elements of hazard management, commencing with the importance of proficient hazard management while examining the historical methodologies and enduring hurdles linked to this critical domain.

7.2.1 DEFINING HAZARDS AND THEIR TYPES

Hazards are situations or occurrences that have the capacity to result in harm, destruction, or negative outcomes. Generally, hazards are grouped into three primary categories:

7.2.2 NATURAL HAZARDS

These hazards originate from naturally occurring processes and include events such as seismic tremors (earthquakes), overflowing bodies of water (floods), powerful tropical storms (hurricanes), violent wind funnels (tornadoes), large-scale forest fires (wildfires), and volcanic eruptions (Perry & Lindell, 2008).

7.2.3 TECHNOLOGY HAZARDS

These hazards are associated with systems and activities created by humans, like accidents in industrial settings, accidental releases of chemicals, incidents involving nuclear facilities, and failure in infrastructure, such as bridges or power grids (Renn, O. 2008).

7.2.4 ENVIRONMENTAL HAZARDS

These hazards emerge as a result of environmental influences, such as the contamination of air and water, the consequences of climate change, and the degradation of ecosystems (Turner *et al.*, 2010).

7.2.5 Importance of Effective Hazard Management

Effectual hazard management is of paramount importance for a variety of reasons. This encompasses:

Human Safety: The fundamental goal of hazard management is to ensure the safety and welfare of people. This means that taking timely and well-suited actions is absolutely essential for minimizing harm and saving lives when hazardous events occur (Cutter, 2008).

Proper Protection: Hazard management aims to minimize damage to property and critical infrastructure. This is crucial for upholding the financial stability of regions and ensuring that communities can quickly bounce back from adversity (Haimes, 2009).

Environmental Stewardship: A comprehensive approach to hazard management involves considering the protection of the environment and its ability to thrive over the long term (Mileti & O'Brien, 1992).

Resilience Building: Efficient hazard management strengthens the ability of communities and regions to endure, react to, and recuperate from hazardous events, making them more robust in the face of adversity (Paton *et al.*, 2001).

It involves a range of activities, including taking steps to mitigate their impacts, improving procedures and plans for responding to hazards when they occur and while they occur in order to minimize damage and save lives, and rebuilding and repairing damage caused by hazards (World Bank, 2014).

7.2.5.1 Historical Approaches and Challenges in Hazard Management

Historically, it often involved reactive responses to disasters. However, the field has transitioned toward a more proactive approach, emphasizing preparedness and community engagement (Hewitt, 1983).

Resource constraints: when resources are scarce, this can hinder various aspects of hazard management, including the ability to prepare for, mitigate, respond to, and recover from hazards (Kapucu, 2008). The data and information challenges of inadequate data and information can make it difficult to make informed decisions in hazard management. Having gaps in what we know can be a stumbling block (Mileti, 1999). The complexity and uncertainty in which hazards are inherently intricate and often accompanied by unpredictability adds to the difficulty of effectively managing them (Smith & Wenger, 2007). The growing trend of urbanization and the impacts of climate change are intensifying the effects of hazards, which means those strategies for managing hazards must evolve to account for these changing conditions (Djalante *et al.*, 2018).

7.3　ELEMENTARIES IN MACHINE LEARNINGS

Machine Learning (ML) is a type of artificial intelligence (AI) that allows computers to learn without being explicitly programmed. ML algorithms can be used to make predictions about new data. The primary objective is to enable computers to generalize from data and apply their knowledge to new, unseen data (Mitchell, 1997). ML is typically categorized into three fundamental learning paradigms:

Supervised Learning: This category of machine learning techniques excels in making predictions and classifications. The algorithms are trained on meticulously curated datasets, where each data point is not only accompanied by its features but also by a designated target that represents the desired outcome. In essence, supervised learning algorithms aim to establish a mapping from input data to known output, enabling them to provide reliable predictions for a wide array of applications, from image recognition to spam email filtering. The model learns to predict the output data for new input data based on the training data.

Unsupervised Learning: This represents a unique realm within the machine learning landscape. It navigates the complex territory of raw, unstructured data space, striving to uncover inherent structure and patterns that may be hidden from plain sight. These methods often serve as the backbone of data preprocessing and feature engineering, including customer segmentation, recommendation systems, and data compression.

Reinforcement Learning: Reinforcement learning takes a fundamentally different approach, resembling the learning process of a curious and adaptive agent in an interactive environment. This type of machine learning involves an algorithm, often referred to as an agent, that engages with its surroundings and learn to make decisions through a series of trials and errors. It is relevant in scenarios like autonomous robotics, game playing, and recommendation systems, where learning by interacting with the environment is more effective than learning from static data. The model learns to maximize its reward over time (Goodfellow *et al.*, 2016).

7.3.1　Key Concepts and Terminologies

A dataset represents a structured collection of data points, each consisting of various attributes. These features can be considered the specific characteristics that describe each data point. A feature, in the context of data analysis and machine learning, is a measurable property of a data point. For instance, in customer data analysis, features may encompass variables like age, gender, purchase history, and any other relevant data that characterizes each customer. The label corresponds to the anticipated or target outcome associated with a data point. A model is the result of training a machine learning algorithm on a dataset. It embodies the algorithm's learned understanding of the patterns, relationships, and associations within the data. Training is the process by which a machine learning algorithm is presented with a dataset to acquire knowledge and insights. A prediction is the outcome generated by a trained machine learning model when given new, previously unseen data. It represents the model's best estimate or inference based on the patterns it has learned from the training data. Accuracy serves as a quantifiable measure of a learning model's performance. It is usually expressed as a percentage and is a key metric in evaluating the effectiveness of a model (Kumar, P. *et al.*, 2021).

7.4 DATA ACQUISITION AND PREPROCESSING

Certainly, in more technical terms, data plays a foundational and indispensable role in hazard management by serving as the primary raw material for advanced information processing and decision-making. This data-driven approach relies on state-of-the-art technology and methods to gather, process, and analyze information. Data is the backbone of hazard management, offering insights and information that enable early warning, risk assessment, and decision-making (Quarantelli, E. L., 1998). Advanced predictive models, powered by big data and artificial intelligence, can anticipate the course of hazards. These models are constructed with machine learning algorithms and deep learning neural networks. In this high-technology landscape, data is the raw material upon which intelligent algorithms, machine learning, and AI processes operate, enabling hazard management to reach unprecedented levels of sophistication and precision (Mileti, D.S., 1999).

Data Sources: (Sensors, Satellites, Social Media, etc.)

Sensors are like high-tech scouts strategically placed in various locations, both nearby and far away. These advanced scouts continuously collect and report information about the environment in real-time. They tell us things like how hot or cold it is, how moist the air is, whether the ground is shaking, or whether the air is safe to breathe. This up-to-the-minute data is vital for keeping tabs on potential hazards and managing them effectively (Turner *et al.*, 2010). In satellites, we can think of satellite-based remote sensing as a high-flying eye in the sky, providing a comprehensive view of our planet. These satellites take pictures and collect information about what's happening on earth from their vantage point in space. They help us understand global changes and potential dangers. The giant cameras in the sky snap pictures of the Earth, showing us things like upcoming weather patterns, how we're using the land like agriculture or cities, if we're cutting down too many trees, and even if natural disasters like wildfires are occurring. This "big picture" perspective from space is like a global monitor that helps us keep track of significant environmental changes and potential risks worldwide (Kuenzer *et al.*, 2014). Throughout and after hazardous events, social media platforms have become valuable sources of data. They provide real-time information and user-generated content that can assist in assessing the impact and response to disasters (Vieweg *et al.*, 2010).

7.4.1 DATA PREPROCESSING TECHNIQUES

In data cleaning, data collected from various sources often contain outliers, errors, or missing values. Techniques in data cleaning, such as imputation, detection, and noise reduction, are applied to enhance data quality and reliability (Dasu & Johnson, 2003). Normalizing data involves scaling it to a standard range, typically between 0 and 1 or with a mean of 0 and standard deviation of 1. This step is crucial for ensuring that features with different scales have equal importance during model training

(Hastie *et al.*, 2009). Feature engineering is the process of selecting, creating, or transforming features to improve model performance. It involves identifying relevant features, reducing dimensionality, and extracting valuable information from the data (Koza, 1996).

7.5 PREDICTIVE MODELS FOR HAZARD FORECASTING

One of the basic applications of ML in intelligent hazard management is hazard forecasting. This division explores predictive models pre-owned for forecasting hazards, focusing on supervised learning approaches, common algorithms, and case studies demonstrating their use. A powerful approach in hazard forecasting is supervised learning, where models are instructed legacy data that includes both accrual character and corresponding hazard outcomes (Cutter, 1996). Some of the common algorithms are regression models, where the goal is to estimate numerical values. The linear and polynomial regression, and more complex variants like Support Vector Regression (SVR), are applied to model kinship of features and hazard outcomes (Hastie *et al.*, 2009). In hazard forecasting the neural networks are mostly used in deep learning techniques. In hazard data, Convolutional Neural Networks (CNNs) and Recurrent Neural Networks (RNNs) can capture complex spatial and temporal patterns (LeCun *et al.*, 2015). Case studies demonstrate the use of Machine Learning for weather, earthquake, and wildfire prediction.

7.6 ANOMALY DETECTION FOR EARLY WARNING

Anomaly detection is a critical aspect of intelligent hazard management, and this section explores the role of anomaly detection for early warning, the requisition of unsupervised learning for anomaly detection, detecting unusual patterns in data, and its application areas like cybersecurity, identifying hazardous events, etc. In unsupervised anomaly detection, the model identifies data points and patterns that deviate notably from the norm without the need for labeled data (Markou & Singh, 2003). Various machine learning and statistical techniques are employed to accomplish this timely response and early warning of emerging hazards (Chandola *et al.*, 2009). In fraud detection, anomaly detection helps to identify unusual transactions that may indicate fraudulent behavior (Bolton *et al.*, 2002). In cybersecurity, it helps to identify suspicious activities within system logs and user behavior for preventing security breaches (Mahmood *et al.*, 2019).

7.7 RISK ASSESSMENT AND MITIGATION

Risk assessment requires the methodical appraisal of the potential impact and probability hazards which helps decisionmakers to understand the level of risk associated with specific hazards and prioritize actions accordingly (Turner *et al.*, 2003). Machine learning is increasingly applied to risk assessment, and models can analyze vast datasets to identify patterns. Risk mitigation and preparedness embrace developing strategies to enhance community resilience and to reduce the impact of hazards.

This includes measures such as infrastructure improvements and public awareness campaigns to prepare for and respond to potential hazards (Paton *et al.*, 2001).

7.8   REAL-TIME HAZARD MONITORING

Real-time hazard monitoring is pivotal for saving lives and property. It helps us quickly and accurately spot dangers and create early warning systems, allowing people to evacuate from risky areas in time. Machine Learning can help create real-time hazard monitoring systems by using data that's generated as it happens, like sensor or social media data (Kumar & Maheshwari, 2021). These systems use algorithms that can learn from this streaming data without needing to keep all the data in memory at once. Fast learning is a challenge in creating computer programs that can swiftly and correctly understand new data as it flows in, so we can spot hazards in real-time. In handling big data, another challenge is building systems that can grow and manage huge amounts of this continuous data effectively (Chapi *et al.*, 2017).

7.9   GEOSPATIAL ANALYSIS AND HAZARD MAPPING

Geospatial data is information tied to particular places on Earth. It helps us study and depict both natural and human-caused dangers. Geographic Information Systems (GIS) are software tools for gathering, storing, and studying geospatial data. We can blend GIS with machine learning to create solutions to manage hazards more effectively (Pradhan *et al.*, 2019). Certainly, in simple terms, spotting risk areas on hazard maps help us find places prone to disasters, and this knowledge guides us in creating plans to respond to emergencies. In resource allocation, it helps us decide where to send help first. For instance, if a map shows areas prone to flooding, we can send resources there promptly, informing the public. They enable us to set up early warnings for dangers and tell people about the best ways to evacuate if needed (United Nations Office for Disaster Risk Reduction UNISDR, 2015).

7.10   HUMAN-CENTRIC APPROACHES AND DECISION SUPPORT

When it comes to intelligently managing hazards, human knowledge and skills are still critical, and this section discusses how experts play a key role, how machine learning aids decision-making, and why it's crucial to have humans overseeing automated systems to make ethical and accurate choices (Paté-Cornell, M.E., 2012). Human experts are vital in managing hazards; they bring knowledge, experience, and understanding of the local situation to make important decisions, interpret data, and ensure a successful response to dangers (Kusiak, 2009). In smart hazard management, being ethical is vital. Human-in-the-loop systems make sure that people are still in charge and keeping an eye on automated processes. This stops us from depending too much on computer algorithms and helps avoid any unfairness or mistakes that might come up (Lepri *et al.*, 2017).

7.11 CASE STUDIES AND APPLICATIONS

In this part, we'll explore detailed real-life examples of how machine learning has been used in different areas of hazard management. These case studies will showcase where it's worked well, the obstacles faced, and the valuable lessons we've gained from these experiences. Machine learning is helping healthcare by forecasting disease outbreaks, spotting illness early, and effectively managing resources. For example, it's used to predict the spread of infectious diseases like COVID-19 (Bansal *et al.*, 2016). In the financial world, machine learning is used to evaluate risks, catch fraudulent activities, and analyze market trends. For instance, it has greatly improved spotting fraud in credit card transactions (Dal Pozzolo *et al.*, 2015). Machine learning is also used for early warnings and keeping an eye on environmental catastrophes. Case studies might delve into how it's used to predict and respond to things like earthquakes, floods, and wildfires (Schmidt *et al.*, 2018). Machine learning in hazard management has achieved more accurate predictions, quicker responses, and better resource use. Problems like unreliable data, understanding how the models work, and making sure they're fair and unbiased are challenges. We've learned that teamwork between different fields is key, getting the right data and preparing it well matters, and we need to keep checking and adjusting our models as things change (Xiong *et al.*, 2015).

7.12 FUTURE TRENDS AND CHALLENGES

As sensors become more advanced and affordable, they will provide a broader range of data, improving the accuracy and depth of hazard assessments in areas like environmental monitoring, healthcare, and infrastructure. Techniques like convolutional neural networks (CNNs) and recurrent neural networks (RNNs) are gaining traction for handling complex spatial and temporal data. They show promise in various hazard prediction and monitoring applications. Privacy concerns drive the adoption of federated learning, where machine learning models are trained using decentralized local data sources, without sharing raw data. This approach preserves data privacy while supporting collaborative model training. The ongoing challenges are data privacy as the collection and sharing of sensitive data raise privacy concerns. Many machine learning models, especially deep learning ones, are considered "black boxes" because their decision is a challenge. As datasets grow in size and complexity, ensuring the efficient scalability of machine learning models becomes crucial. Addressing biases inherited from training data in machine learning models is an ongoing concern to prevent unfair or discriminatory outcomes. Adapting to these emerging trends and effectively addressing these ongoing challenges is vital in shaping the future of machine learning in hazard management, ensuring that intelligent systems can efficiently and ethically protect communities from various hazards.

7.13 CONCLUSION

The marriage of machine learning and intelligent hazard management represents a dynamic and ever-evolving partnership that holds immense potential for creating safer, more resilient communities. By embracing these technologies, interdisciplinary collaboration, ethical considerations, and innovative research, we are poised to build a future where machine learning plays a pivotal role in protecting our world from a multitude of hazards.

REFERENCES

Bansal, S., Preisler, B. A., Brewer, N. V., Dredze, M. E., Hedlin, M. H., Johnson, T. M., Johnson, J. M., Meltzer, C. T., & Paul, D. S. (2016). "Prediction and early warning of diseases using the internet and social media". Online Journal of Public Health Informatics, 8(2).

Bolton, R. J., & Hand, D. J. (2002). "Statistical fraud detection: A review". Statistical Science, 17(3), 235–249.

Chandola, V., Banerjee, A., & Kumar, V. (2009). "Anomaly detection: A survey". ACM Computing Surveys (CSUR), 41(3), 1–58.

Chapi, K., Singh, V. P., Shirzadi, A., & Tiefenbacher, J. (2017) "A machine learning approach for flood susceptibility mapping, using multi sensor satellite data and GIS". Water, 9(10), 738.

Cutter, S. L. (1996). "Vulnerability to environmental hazards". Progress in Human Geography, 20(4), 529–539.

Cutter, S. L., &Boruff, B. J. (2008). "A place-based model for understanding community resilience to natural disasters", Global Environmental Change, 18(4), 598–606.

Dal Pozzolo, A., Caelen, C., Alsharnouby, R., Johnson, T., & Bontempi, G. (2015). "Calibrating probability with under sampling for unbalanced classification". In International Conference on Applications of Natural Language to Information Systems, 33–48.

Dasu, T., & Johnson, T. (2003). Exploratory Data Mining and Data Mining. Wiley.

Djalante, R., Holley, M., Thomalla, A., Carnegie, J., & Lassa, I. (2018). Urbanization and climate change risks: A global challenge with local implications. Environmental Science & Policy, 80, 84–93.

Gaber, M. M., Zaslavsky, A., & Krishnaswamy, S. (2005). "Mining data streams: A review". SIGMOD Record, 34(2), 18–26.

Goodfellow, I., Bengio, Y., & Courville, A. (2016). Deep Learning. MIT press.

Haimes, Y. Y. (2009). "On the complex definition of risk: A systems-based approach". Risk Analysis, 29(12), 1647–1654.

Hastie, T., Tibshirani, R., & Friedman, J. (2009). The Elements of Statistical Learning: Data Mining, Inference, and Prediction. Springer.

Hewitt, K. (1983). The idea of calamity in a technocratic age. In D. J. Pogue (Ed.), Disasters, Collective Behavior, and Social Organization (pp. 1–38). Lexington Books.

Kapucu, N. (2008). Collaborative decision-making in emergency and disaster management. International Journal of Public Administration, 31(3), 279–290.

Koza, J. R. (1996). "Automated discovery of submodels in data". Artificial Life, 2(4), 323–331.

Kuenzer, C., Dech, A. C., Eineder, H. E., & Dech, S. (2014). "Remote sensing for assessing natural hazards". Progress in Physical Geography, 38(2), 153–163.

Kumar, P., & Maheshwari, P. K. (2021). "Machine learning models for intelligent hazard management: A review of the literature". International Journal of Disaster Risk Reduction, 73, 102894.

Kusiak, Andrew (2009). "Innovation: A data-driven approach," International Journal of Production Economics, Elsevier, vol. 122(1), pages 440–448, November.

LeCun, Y., Bengio, Y., & Hinton, G. (2015). "Deep learning". Nature, 521(7553), 436–444.

Lepri, B., Andrienko, N., Boldrini, C., Conti, M., De Munari, I., Blanco, R., & Marquardt, N. (2017). "The tyranny of data? The bright and dark sides of data-driven decision-making for social good". In Proceeding of the 2017 ACM Conference on Computer Supported Cooperative Work and Social Computing, 767–780.

Mahmood, A. N., Hu, H., & Poor, H. V. (2019). "Machine learning for cyber physical systems security: A survey". IEEE Access, 7, 183225–183253.

Markou, M., & Singh, S. (2003). "Novelty detection: A review-part 1: Statistical approaches". Signal Processing, 83(12), 2481–2497.

Mileti, D. S. (1999). Disasters By Design: A Reassessment of Natural Hazards in the United States. Joseph Henry Press.

Miletti, D. S., & O' Brien, P.W. (1992). "Warnings during disasters: Normalizing communicated risk". Social Problems, 39(1), 40–57.

Mitchell, T. M. (1997). Machine Learning. McGraw-Hill.

Paté-Cornell, M. E. (2012). " 'On Black Swans' and 'perfect storms': Risk analysis and management when statistics are not enough". Risk Analysis, 32(11), 1823–1833.

Paton, D., Johnston, D., & Houghton, B. (2001). "Information in practice: Household preparedness for natural hazards: Impact of neighborhood density and personal responsibility". Environment and Behavior, 33(4), 472–489.

Paton, D., McClure, J., Johnston, D., Houghton, C., & Mclvor, R. G. H. (2008). "Community resilience: Integrating hazard, risk and social vulnerability". Disaster Prevention and Management, 17(1), 270–277.

Perry, R. W., & Lindell, M. K. (2008). Understanding and managing public response to disaster: Lessons from four case studies. Environment Hazards, 8(2), 81–95.

Pradan, B., Sezer, E. A., Gokceoglu & Pradhan, S. (2019). Machine learning methods in landslide hazard assessment: A review. Computers & Geosciences, 135, 109–121.

Quarantelli, E. L. (1998). "What is a disaster recovery: Operationalizing an existing agenda". Policy Sciences, 40(2), 131–146.

Renn, O. (2008). Risk Governance: Coping with Uncertainty in a Complex World. Earthscan.

Schmidt, D. A., Dresp, B., Franke, J., Conrad, C., Naumann, M., Schulz, K-P., Kresse, W., Wissel, B., & Fritz, J.-S. (2018). "A survey of big data architectures and machine learning algorithms in large-scale environmental monitoring". Sensors, 18(11), 396.

Smith, D. I., & Wenger, D. (2007). "Sustainable disaster recovery: Operationalizing an existing agenda". Policy Sciences, 40(2), 131–146.

Turner, W. R., Spector, S. E., Leidner, N., Rondinini, A., Brooks, T. M., Gascoigne, C., Hayes, R. A., Mittermeier, C. H., Morrison, J., Schipper, R. B., & Butchart, S. H. M. (2010). "Free and open-access satellite data are key to biodiversity conservation". Biological Conservation, 173, 173–176.

Turner BL 2nd, Kasperson RE, Matson PA, McCarthy JJ, Corell RW, Christensen L, Eckley N, Kasperson JX, Luers A, Martello ML, Polsky C, Pulsipher A, Schiller A. A framework for vulnerability analysis in sustainability science. Proc Natl Acad Sci USA. 2003 Jul 8;100(14):8074–9. doi: 10.1073/pnas.1231335100. Epub 2003 Jun 5. PMID: 12792023; PMCID: PMC166184.

United Nations Office for Disaster Risk Reduction (UNISDR). (2015), Global Assessment Report on Disaster Risk Reduction. UNISDR.

Vaya', E. J., S'anchez, O., Zorrilla, E., & Zorrilla, A. (2013). "Artificial neural networks applied to the prediction of forest fires". Expert Systems with Applications, 40(10), 4028–4035.

Vieweg, S., Palen, L., Sutton, J., Liu, S. (2010). "Microblogging during two natural hazards events: What Twitter may contribute to situational awareness". In Proceedings of the SIGCHI Conference on Human Factors in Computing Systems (CHI '10), 1079–1088.

World Bank. (2014). Disaster Risk Reduction: A New Model for an Old Challenge. World Bank.

Xiong, L., Zhang, D., Chen, L., & Pei, J. (2015). "Incorporating domain knowledge into medical diagnosis: An application in predicting kidney disease". In Pacific-Asia Conference on Knowledge Discovery and Data Mining, 147–159.

8 Optimal Dispatch of Distributed Renewable Energy Sources in Isolated Microgrid System Exploiting Metaheuristic Optimization Algorithms

Anirudh Kumar Verma, Prakash M and Shakila B

8.1 INTRODUCTION

Renewable energy sources are fundamentally unpredictable. The optimal utilisation of available resources to fulfil increased electricity demand showed the need for renewable energy sources to be integrated. A microgrid is a programmable entity made up of interlinked loads and distributed energy supplies. The differences between conventional power networks and microgrids are significant: the yielded abilities to generate electricity from micro-sources are substantially lower when compared to what a conventional power plant can produce. Microgrid enterprises are often closer to the client mass, resulting in lower transmission line damage. It is more reasonable for a microgrid to supply energy to remote regions, while it is nearly impractical for the national lattice. Microgrids, when adequately incorporated, can help provide the adaptability and reliability required to fulfil the demand for a steady supply of electricity in the foreseeable future. Most of the time, individual microgrids will function in a grid-tied manner, with supply flowing both ways between the microgrid and the adjacent network. A paralleled bilateral link can help meet functional objectives including enhanced reliability, cost savings and energy source heterogeneity. The ability to operate independently of the grid enhances reliability and enables a standby or emergency mode of functioning [1].

Microgrids can be run in synchrony along the electrical grid as well as with stand-alone off-grids. Numerous research groups have undertaken extensive research to identify the significance and benefits of microgrids. Even while the underlying principles

DOI: 10.1201/9781003452393-8"

of microgrids are well understood, their application, particularly in the case of remote microgrids, is not always equally understood. So much research has gone into grid-connected microgrids that there is still a lot of room for isolated MG research. Furthermore, the cost-effectiveness of the MG system setup is a major concern for real-time implementation.

A battery-powered energy storage system minimises the microgrid's fuel costs by retaining extra power from non-conventional renewable resources and transferring it into the electrical grid as needed. The working status of the generating units will be defined by Unit Commitment. The appropriate allocation of power requirements across committed generators is critical for lowering overall fuel costs associated with meeting load demand [2].

As of now, a considerable amount of research is being performed around MGs; for example, an overview of microgrid issues as well as research findings in areas such as distributed generation, MG economics, power electronics applications, MG operation and control, MG clustering, and protection and communication concerns is presented by Sina Parhizi et al. [3]. Economic consideration for a PV and Energy storage system hybrid grid connected MG using few metaheuristic algorithms has been proposed in [4]. An efficient planning for isolated MGs using hybrid AC/DC topology and metaheuristic algorithms is presented in [5]. For these purposes, biogas generators are studied alongside prominent renewable sources such as wind generators and photovoltaics for a certain yearly load profile. PSO, ABC, CS and PSO-CS Hybrid algorithms were implemented and investigated to acquire the minimum TCOI.

8.2 PROPOSED MICROGRID SYSTEM DESIGN

A proposed remote MG system comprises WTs, PVs, PV controllers, batteries, BGs, inverters and demanded load.

8.2.1 PV Model

Climatic conditions have a huge impact on the output of a PV system. The National Aeronautics and Space Administration (NASA) provides data on solar radiation. Total energy generated per year (kWh/yr) and output power (kW) from a PV array for a specific location can be estimated by the equations (1) and (2), respectively [6]:

$$E_{PVS} = T_{yrhrs} \sum_{G_{min},\ T_{min}}^{G_{max},\ T_{max}} P_{PVS} \tag{1}$$

$$P_{PVS} = P_{R\_STC}\, D_F\, \frac{G_F}{G_{F,STC}}\,(1 + \theta\,(T_{cell} - T_{ref})) \tag{2}$$

where P_{PVS} is the power output from the PV array and T_{yrhrs} represents the total hours in a year. P_{R_STC} is the maximum capacity of the PV system (W) at standard test conditions, D_F is the degrading factor (%), G_F represents solar irradiation falling on the PV system (kW/m^2) and $G_{F,STC}$ represents solar irradiation falling on PV system (kW/m^2) at standard test conditions STC, T_{cell} and T_{ref} are the PV system temperature and STC temperature (in °C) respectively and θ is the temperature coefficient of P_{PV} and D_F.

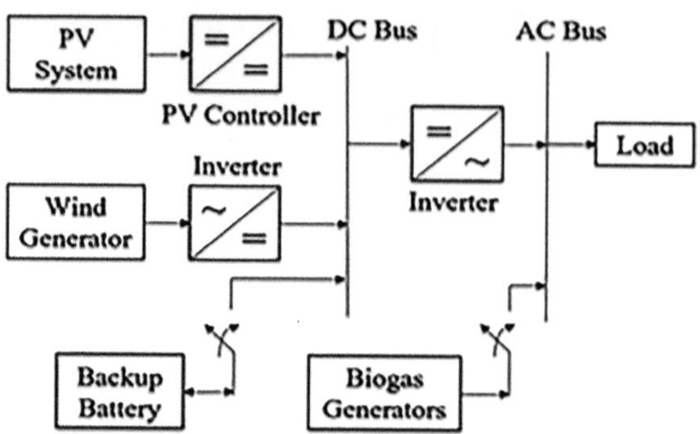

FIGURE 8.1 Diagram of the proposed microgrid.

8.2.2 WIND GENERATION MODEL

The output of a wind generator with a known shape and scale factor can be calculated by the average wind speed for a certain location at a given height using NASA's surface meteorological and solar energy data. The measured wind speed must be adjusted according to the hub height (h_{hub}), which can be achieved using wind data from NASA's database using equation (3) [7]:

$$v = v_{refer}\left(\frac{h_{hub}}{h_{refer}}\right)\beta \tag{3}$$

where v is the wind speed at the desired height, h_{refer} is the reference height, h_{hub} is the hub height and β is the shear exponent coefficient. The generally used value for β is 1/7. Total Energy generated per year (kWh/yr) from a wind generator is estimated by the following equation (4): [7]

$$E_{WG} = T_{yrhrs} \sum_{V_{min}}^{V_{max}} P_o f(v,\psi,\rho) \tag{4}$$

where T_{yrhrs} are the hours in a year, (v_{min}, v_{max}) represents the minimum and maximum wind speed, P_o is the power output from the wind generator and $f(v,\psi,\rho)$ indicates the Weibull probability distribution function with scale parameter ψ and shape parameter ρ.

8.2.3 BIOGAS GENERATION MODEL

Animal excrement, such as that of cattle, goats, horses, donkeys and mules, can be used to determine biogas potential. Total Energy produced per day by the biogas generator (kWh/day) is given by [8]:

$$E_{BIO} = (T_G * CV_{BIO} * \eta_{DieselEng} * \eta_{biogas}) / 860 \tag{5}$$

where T_G is the Total biogas produced at a location (Kg/day) considering 70% collection efficiency potential and biogas availability as 0.019% and CV_{Bio} is the Calorific value of biogas (4700 kcal/m³), $\eta_{DieselEng}$ is Diesel engine efficiency considered equal to 25% and η_{biogas} is biogas generator efficiency considered equal to 95% [9, 10].

8.2.4 BATTERY BANK MODEL

Due to the intermittency of solar and wind generators, energy backup storage is required to support MG. The capacity of battery backup (Ah) required is given as (6) [11].

$$Q_{BB} = \frac{ACL_{daily} * N_{auto}}{D_{dis}} \tag{6}$$

$$ACL_{daily} = E_{Daily} / V_{sys} \tag{7}$$

where ACL_{daily} is the Ah utilisation of daily load demand, E_{Daily} daily load demand in kWh, D_{dis} is the depth of discharge and N_{auto} is the autonomous day.

8.2.5 INVERTER MODEL

For remote MGs, standalone inverters are ideal. The number of required inverters can be evaluated using the following Eq. (12):

$$N_{IVR} = P_{Gen,max} / P_{IVR,max} \tag{8}$$

where $P_{Gen,max}$ is the maximum power (kW) generated by the DERs of MG and $P_{IVR,max}$ is the maximum capacity that can be produced by the inverter.

8.2.6 PV CONTROLLER MODEL

The MPPT controller is employed as a PV controller in this study, and it tracks the PV array's maximum power point during the day to supply the maximum solar irradiation to the system. The number of controllers (N_{CNR}) required for a PV array system is determined by using the following equations:

$$P_{PVS,max} = N_{PVS} * P_{PVS,R} \tag{9}$$

$$P_{CNR,max} = V_B * I_{CNR} \tag{10}$$

$$N_{CNR} = P_{PVS,R} / P_{CNR,max} \tag{11}$$

where $P_{PVS,max}$ is the rated power of the PV system at STC, N_{PVS} is the total number of PV arrays, I_{CNR} represents the maximum current that can be handled by the controller from PV array to battery and N_{CNR} indicates the number of required controllers.

8.3 OPTIMAL SIZING PROBLEM FORMULATION

The optimal dispatch problem was created to reduce the total cost of the microgrid while satisfying all constraints.

8.3.1 OBJECTIVE FUNCTION

To lower TCOI, the number of wind turbines, diesel generators, PV modules, controller units, batteries and inverter units should all be appropriately selected. The techno-economic information on the commercial components are utilised in this investigation [10].

$$Minimize \quad TCOI = \sum_{k=1}^{n_{PVSk}} N_{PVSk} C_{PVSk} + \sum_{l=1}^{n_{WGl}} N_{WGl} C_{WGl}$$

$$+ \sum_{m=1}^{n_{BIOm}} N_{BIOGENm} C_{BIOGENm} + \sum_{p=1}^{n_{BBp}} N_{BBp} C_{BBp} \qquad (12)$$

$$+ \sum_{q=1}^{n_{PVSq}} N_{CNRq} C_{CNRq} + \sum_{r=1}^{n_{IVRr}} N_{IVRr} C_{IVRr}$$

where TCOI = total cost of Investment, N_{WG} = number of wind turbines, N_{PVS} = PV modules, N_{BB} = batteries, N_{CON} = controller units, N_{INV} = inverter units and N_{BIOGEN} = bio generator units, respectively. C_{WG}, C_{PVS}, C_{BIOGEN}, C_{BB}, C_{CNR} and C_{IVR} represent the TCOIs of the wind generator, PV system, bio generator, battery, controller and inverter units respectively, and their values can be evaluated using the following set of equations:

$$Wind\ Generator\ cost: C_{WG} = C_{CapWG} + C_{InsWG} + T_{Life}\ C_{onmWG} \qquad (13)$$

$$PV\ system\ cost: C_{PV} = C_{CapPVS} + C_{InsPVS} + T_{Life} C_{onmPVS} \qquad (14)$$

$$Battery\ Backup\ cost: C_{BB} = C_{CapBB} + C_{InsBB} + C_{RepBB} N_{RepBB} \qquad (15)$$

$$Cost\ of\ Inverter: C_{IVR} = C_{CapIVR} + C_{InsIVR} + C_{RepIVR} N_{RepIVR} + C_{onmIVR} T_{Life} \qquad (16)$$

$$Controller\ cost: C_{CNR} = C_{CapCNR} + C_{InsCNR} + C_{RepCNR} N_{RepCNR} + C_{conmCNR} T_{Life} \qquad (17)$$

$$Biogas\ Generator\ cost: C_{BIOGEN} = C_{CapBIOGEN} + C_{InsBIOGEN} + T_{Life} C_{onmBIOGEN} \qquad (18)$$

8.3.2 CONSTRAINTS

The objective function's equality constraint (energy balance equation) and inequality constraints are shown in the following equations:

$$\sum_{k} N_{WG} E_{WG} + \sum_{l} N_{PVS} E_{PVS} + \sum_{m} N_{BIOGEN} E_{BIOGEN} \geq \frac{E_{Dem}}{y_{MG}} \qquad (19)$$

$$\eta_{MG} = \eta_{BB}\ \eta_{Con}\ \eta_{inv}\ \eta_{W} \qquad (20)$$

where E_{Dem} = energy demand, E_{WG} = *energy generated from wind*, E_{PVS} = energy generated from *PV* modules, E_{BIOGEN} = energy generated from bio-generators, η_{system} = overall efficiency of the MG system, η_{inv} = efficiency of inverter, η_{sys} = efficiency of PV controller, η_w = efficiency of connection wires, η_{BB} = efficiency of battery, SOC = state of charge of the batteries, P_{IVR} = output of the inverter, P_{CNR} = output of the controller and P_{max} = maximum energy generated in MG.

8.3.3 ARTIFICIAL BEE COLONY ALGORITHM

Karaboga proposed the Artificial Bee Colony (ABC) in 2005, which is based on honeybee swarms' intelligent foraging behavior. Honeybees hunt for the ideal nest site among many options, balancing speed and accuracy. ABC was created by observing the behavior of real bees when looking for nectar and sharing information about food sources with the other bees in the hive. A strong example of a swarm-based decision procedure is the group decision-making process utilised by bees to find the best food resources among many solutions. Algorithm steps are mentioned here [12]:

Initialising Step: The initial sources of food are provided at random using the equation:

$$z_m = lb_i + rand\,(0,1)*(ub_i - lb_i) \tag{21}$$

where ub_i = upper bound, lb_i = lower bound, *rand (0, 1)* = random number within the range [0–1].

Employed Bee Phase: The neighbour food source can be evaluated as

$$o_{mi} = z_{mi} + \theta_{mi}(z_{mi} - _{ki}) \tag{22}$$

where i = parametre index, z_k = food source, θ_{mi} = random number ranging from [−1, 1].

Onlooker Bee Phase: The quantity of a food source is evaluated by its profitability and the profitability of all food sources. $Prof_m$ is determined by the formula:

$$Prof_m = fit_m(z_m) / \sum_m fit_m(z_m) \tag{23}$$

where $fit_m(z_m)$ = fitness of z_m.

Scout Phase: The new output is randomly searched by the scout bees. The new output of z_m will be discovered by the scout by using the expression:

$$z_m = l_i + rand\,(0,1)*(ub_i - lb_i) \tag{24}$$

where *rand (0, 1)* = random number ranging from [0, 1], ub_i = upper bound and lb_i = lower bound.

8.3.4 CUCKOO SEARCH OPTIMISATION

Cuckoo Optimisation was created by Yang and Deb in 2009, while Rajabioun created the Cuckoo Optimisation Algorithm in 2011. The cuckoo search was first presented as a technique for mathematical optimisation problems and continuous problems, based on the nesting parasitism of cuckoo birds. Studies performed this algorithm to the test on several well-known test functions, comparing it to PSO and GA, and found that cuckoo search outperformed PSO and GA. Since then, the algorithm's creators and numerous researchers have applied it to optimisation problems with encouraging results. Algorithm implementation can be done using the equations that follow [11]. The formation of initial nests of swarm birds:

$$N_{i,k}^0 = round(z_{k,min} + rand(z_{k,max} - z_{k,min})) \tag{25}$$

where $N_{i,k}^0$ is the variable's initial value for the nest, $(z_{k,min}, z_{k,max})$ are the variable's minimum and maximum permissible values and is a variable parametre having range [0, 1]. Due to the discrete structure of the optimisation issue, the function is used for generation of new cuckoo nests using Lévy flight. Except for the best, all of the nests are relocated based on the quality of new cuckoo nests established by Lévy flights from their original locations as specified.

$$N_i^{t+1} = N_i^t + \lambda.step.(N_i^t - N_b^t).random \tag{26}$$

where N_i^t denotes the present position of the nest, $randnum$ is a random number which is normally distributed and λ is the step size factor. The position of the best nest, N_b^t, is determined via a random walk based on Lévy flight. Step length:

$$S_{len} = \frac{u}{|v|^{1/\beta}} \tag{27}$$

where β is a chosen between (1, 2) and (u, v) and is obtained from normal distribution:

$$\sigma_u = \left\{ \frac{\Gamma(1+\beta).sin(\beta_\pi - 2)}{\Gamma[(1+\beta)/2].\beta.2^{(\beta-1)/2}} \right\} 1/\beta \tag{28}$$

where $\Gamma(z)$ represents the gamma function and $\beta \in (1,2)$.

For each solution, the alien eggs are discovered using a probability $matrix.p_{ik} =$

$$\begin{cases} 1; & randnum < p_d \\ 0; & randnum \geq p_d \end{cases} \begin{matrix} (29) \\ (30) \end{matrix}$$

where the probability of discovery is p_d. Through random walks with the following step size, fresh created eggs of good quality will replace existing eggs from their current places:

$$S_{len} = rand\,(N(randp_1(n),:) - N(randp_2(n),:)) \tag{31}$$

$$N^{t+1} = N^t + S_{len} * P_m \tag{32}$$

where *randp* is a random permutation function used for different rows' permutation applied on the nests' matrix, and P_m is the probability matrix. The steps used to discover the alien eggs and generate new cuckoos are alternatively performed until a termination criterion is reached.

8.3.5 PSO-CS Algorithm

PSO-CS is a 2019 proposed algorithm [11] that integrates a swarm-based algorithm (particle swarm optimisation) with a bio-inspired algorithm (cuckoo search). Initial working of the algorithm till the updating and weight factor part is based on PSO [13, 14] using equations (29) and (31) respectively. A Lévy flight approach is used for updation of velocity, which can be illustrated from Eq. (52):

$$v_{i,k+1} = W_k v_{i,k} + C_1 \oplus L(\tau)(lb_{i,k} - X_{i,k}) + C_2 \oplus L(\tau)(gb_{g,k}) - X_{i,k}) \tag{33}$$

where $v_{i,k+1}, W_k v_{i,k}, C_1, lb_{i,k}, X_{i,k}, C_2, gb_{g,k}, X_{i,k}$ represents same as equation (32) while $L(\tau)$ represents Lévy flight parametre and $\oplus$ represents entry wise multiplication.

8.4 RESULTS AND DISCUSSIONS

The TCOI is evaluated using a yearly load profile over a 5-year period, with a project life of 20 years. The amount of generating components required for the estimated loading on the units based on simulation can be determined in a way that is retained within defined restrictions whilst also satisfying the load requirement. A collection of batteries has also been used to store extra energy if the generation exceeds the load, and to deliver electricity if the load exceeds the generation. Inverter capacity is determined by the quantity of batteries, while controller capacity is determined by the PV generation. The Artificial Bee Colony (ABC) Optimisation Algorithm, Cuckoo Search Optimisation Algorithm, and Particle Swarm-Cuckoo Search Hybrid Algorithm have been used in the simulation, which was conducted in MATLAB. Following that, the outcomes for various parameters are compared. The MG components' techno-economic data is taken from [11].

There will only be one optimal solution, and the solution space would have more discontinuities because of the convex character of the objective functions and constraints. By precisely exploring the entire solution space, it is possible to avoid local optima stalling. Convergence graphs for the producing unit's TCOIs are shown in the following. In scenario 2, the ABC optimisation algorithm was applied as shown in Figure 8.2, and the simulation results for ABC show fast convergence, as shown in Figure 8.2, avoiding falling into local minima due to the scouting phase. The TCOI for generating sources is \$ 277,876.57, while the TCOI for the entire MG system is \$ 322,018.74.

Scenario 3 comprises the implementation of the CSOA approach, and unlike PSO and ABC simulation outcomes depicted in Figure 8.2, CSOA simulation results reveal a slower convergence rate but a larger TCOI for producing sources (\$ 269,317.99), which is slightly higher than the PSO-CS hybrid approach, while the TCOI for the entire MG system was \$ 269,317.99, which was the least of all the approaches used. The probability matrix approach could be the reason for the slower convergence rate. ABC is fast

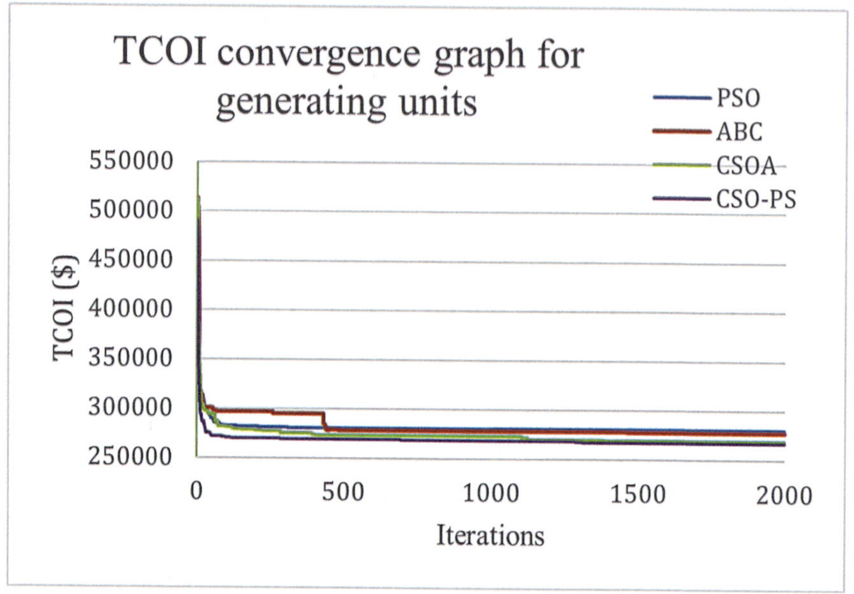

FIGURE 8.2　TCOI convergence curves for generating units.

converging while CSOA was slow converging, but gave better cost results for dispatching parameters, i.e., for generation while TCOI for the overall MG was slightly higher than the classic CSOA. TCOI for generating sources obtained was minimum $ 267,417.35 which is slightly less than the CSOA approach whilst suggesting a slightly higher value of TCOI than CSOA. The TCOI for the entire MG system is $ 313,919.02, indicating that the classical CSOA approach yielded optimal results for the current study.

Table 8.1 shows that if an optimisation method has a better balance between the exploration and exploitation phases, the outcomes are substantially better. Optimal dispatching for such an MG system is established using the metaheuristic algorithms approach for MG components. Four different metaheuristic algorithms were implemented amongst which paper proposes CSOA to determine the optimal minimal total cost of investment for the proposed isolated microgrid. An improvement of 2.59% in not well defined between exploration and exploitation, its performance lags behind that of other optimisation techniques.

8.5　CONCLUSION

The current study established an isolated microgrid system incorporating various microgrid components including wind generators, photovoltaic systems, biogas generators, batteries, controllers and inverters. All the components were modelled and system boundaries and constraints were discovered. The loading on different generating sources in TCOI can be observed using the CSOA technique when compared to PSO.

TABLE 8.1

Comparison of Results for Different Algorithms

Optimization Technique	PSO[1]	ABC	CSOA	PSO-CS
Generated Energy (kWh/yr)	323642.13	351922.60	335227.92	335082.43
Surplus Energy (kWh/yr)	2038.24	30318.72	13624.03	13129.05
Wind Generation (kWh/yr)	436.99	5649.57	5166.30	5422.16
PV Generation (kWh/yr)	165525.14	188593.03	224941.61	224540.27
BG Generation (kWh/yr)	157680.00	157680.00	105120.00	105120.00
TCOI for Generating units ($)	280913.48	277876.57	269317.99	267417.35
TCOI of MG ($)	322266.41	322018.74	311811.40	313919.02

REFERENCES

1. D. Ton and J. Reilly, "Microgrid Controller Initiatives: An Overview of R&D by the U.S. Department of Energy," *IEEE Power and Energy Magazine*, vol. 15, no. 4, pp. 24–31, 2017.
2. P. K. Mohan Reddy and M. Prakash, "Optimal Dispatch of Energy Resources in an Isolated Micro-Grid with Battery Energy Storage System," 2020 4th International Conference on *Intelligent Computing and Control Systems (ICICCS)*, pp. 730–735, 2020.
3. Md M. Islam, M. Nagrial, J. Rizk, and A. Hellany, "General Aspects, Islanding Detection, and Energy Management in Microgrids: A Review," *Sustainability*, vol. 13, no. 16, p. 9301, 2021.
4. X. Han, H. Zhang, X. Yu, and L. Wang, "Economic Evaluation of Grid-connected Micro-grid System with Photovoltaic and Energy Storage under Different Investment and Financing Models," *Applied Energy*, vol. 184(C), pp. 103–118, 2016.
5. B. Hernandez-Ocana, J. Hernandez-Torruco, O. Chavez-Bosquez, M. B. Calva-Yanez, and E. A. Portilla-Flores, "Bacterial Foraging-Based Algorithm for Optimizing the Power Generation of an Isolated Microgrid". *Applied Science*, vol. 9, p. 1261, 2019.
6. A. Haffaf, F. Lakdja, R. El Meziane, and D. Ould Abdeslam, "Study of Economic and Sustainable Energy Supply for Water Irrigation System (WIS)." *Sustainable Energy, Grids and Networks*, vol. 25, 2021.
7. R. Younis, D. Ibrahim, E. Aboul-Zahab, and A. El'Gharably, "Techno-Economic Investigation Using Several Metaheuristic Algorithms for Optimal Sizing of Stand-Alone Microgrid Incorporating Hybrid Renewable Energy Sources and Hybrid Energy Storage System", *International Journal on Energy Conversion (IRECON)*, vol. 8, no. 4, pp. 141–152, 2020.
8. A. Bhatt, M. Pal Sharma, and R. Saini, "Feasibility and Sensitivity Analysis of an Off-grid Micro hydro–photovoltaic–biomass and Biogas–diesel—battery Hybrid Energy System for a Remote Area in Uttarakhand state, India," *Renewable & Sustainable Energy Reviews*, vol. 61, pp. 53–69, 2016.
9. M. Elfeki, E. El-Bestawy, and E. Tkadlec, "Bioconversion of Egypt's Agricultural Wastes into Biogas and Compost," *Polish Journal of Environmental Studies*, vol. 26, 2017. DOI: 10.15244/pjoes/69938
10. M. Elfeki and E. Tkadlec, "Treatment of Municipal Organic Solid Waste in Egypt," *Journal of Materials and Environmental Science*, vol. 6, pp. 756–764, 2015.

11. J. Ding, Q. Wang, Q. Zhang, Q. Ye and Y. Ma, "A Hybrid Particle Swarm Optimization-Cuckoo Search Algorithm and Its Engineering Applications", *Mathematical Problems in Engineering*, vol. 2019, Article ID 5213759, 12 pages, 2019. https://doi.org/10.1155/2019/5213759
12. A. Sharma, A. Sharma, S. Choudhary, R. Kumar Pachauri, A. Shrivastava and D. Kumar, "A Review on Artificial Bee Colony and It's Engineering Applications," *Journal of Critical Reviews*, vol. 7, no. 11, 2020.
13. M. Prakash and P. Chinnamuthu, "Techno-Economic Modeling of Integrated Renewable Energy System using Adaptive Inertia Weight based PSO," *NISCAIR-CSIR*, vol. 79, pp. 647–652, 2020
14. M. Prakash and P. Chinnamuthu, "Optimal Sizing of Integrated Renewable Energy System Based on Demand Response," IEEE International Conference on Advances in Computing and Communication Engineering (ICACCE), pp. 35–40, June 2018.

9 Leveraging Machine Learning Models for Intelligent Hazard Management

Preethi Nanjundan, George Jossy P, and Karpagam C

9.1 INTRODUCTION

Over the last 15 years, people in the majority of developed nations have grown increasingly aware of the risks associated with technology, established new organizations and initiatives to regulate it, and fundamentally altered the processes involved in the development, manufacturing, and application of technology. More subtle threats will come our way during the next fifteen years, tensions and inconsistencies will surface in the new institutions, and we will continue to be astounded by the bizarre ways that our technology inadvertently harms us [1].

As the name implies, hazard identification is the process of identifying and characterizing dangers that have affected or may affect a focal region. Numerous techniques, such as historical research, ideation, scientific analysis, and subject area expertise, can be used to accomplish this. The following four steps are present in nearly all HRM approaches, regardless of the procedures employed: [2]

- Determine the risks.
- Analyze the hazards associated with each listed danger.
- Examine how the various hazards relate to one another.
- Handle the risk of hazards according to priority.

Create a hazard profile to evaluate and examine a hazard more thoroughly. Profiles are brief reports that include important details about the hazard, its existence in the region of concern, and its nature. Examples of information that is frequently investigated are the following: [2]

- A summary of the hazards' general orientation;
- The location of the hazards within and around the study area, as well as the spatial extent of their effects;
- The length of an event that the hazards causes; seasonal or other time-based patterns that the hazards follow;

DOI: 10.1201/9781003452393-9

- The speed at which an actual hazard event begins; and
- The availability of warnings for the hazards

Organizations face a wide range of risks in the fast-paced, globally connected world of today, risks that can affect their operations, reputation, and long-term performance. With its capacity for transformation, technology is now a crucial enabler for proactive and efficient risk management [3].

9.2 RECENT PROGRESS AND APPLICATIONS IN MACHINE LEARNING

Various machine learning and deep learning techniques are employed to support disaster management throughout all stages. Support vector machines (SVM), Naïve Bayes (NB) techniques, decision trees (DT), random forests (RF), logistic regression (LR), and the K-nearest neighbor (KNN) clustering algorithm are examples of machine learning (ML) techniques. Conversely, deep learning techniques encompass various ANN architectures, including transformer architecture, generative adversarial networks (GANs), recurrent neural networks (RNNs), long short-term memory neural networks (LSTM), convolutional neural networks (CNNs), and multi-layer perceptrons (MLP) [4].

Large and complicated datasets can be used with the help of machine learning and deep learning to create systems that can anticipate disasters, aid in their reaction and recovery, and produce useful decision-support tools. These methods make use of the capacity to work with many forms of data from various sources and spot patterns that can reveal intelligence that would otherwise be impossible to reveal. Unmanned aerial vehicles (UAVs), wireless sensor networks, social media, crowdsourcing, geographic information systems (GIS), and satellite imagery are some of the sources of big data [4].

- Nowadays, in the era of various other emerging technologies, such as unmanned aerial vehicles (UAV), Internet of Things (IoT), and satellite-based technology, the network is becoming more autonomous. Such systems require several local decisions to be made, such as bandwidth selection, data rate selection, power control, and user association to a base station. We can use ML algorithms to address these issues and lower human intervention in uncertain and stochastic environments. To summarize, ML algorithms have the following advantages over other existing technologies [5].
- Large amounts of data are quickly processed by ML algorithms, which can then be used to spot trends. Additionally, the ML algorithms make it simple to analyze different kinds of data. The popularity of machine learning has also grown due to its everyday applications in areas like online customer service, video surveillance, and traffic forecasting [5].
- Fake message detection can be aided by machine learning systems that rely on rules. ML algorithms minimize the requirement for human decision-making and intervention. These methods aid in the suppression of rumors, particularly in the event of man-made disasters [5].

- As the amount of data grows, machine learning algorithms typically perform better. For instance, the predictive power of an earthquake prediction model rises with the amount of data [5].
- Multidimensional data may be handled by ML algorithms, and they can identify outliers in the dataset. Outlier analysis is a crucial method when dealing with high threats. When we are attempting to predict extremely rare events like disasters or pandemics, we should pay particular attention to these outliers rather than eliminating them entirely [5].
- Various machine learning techniques can be used in these situations to make decisions quickly and accurately. This chapter's subsequent subsections detail how these algorithms can be used to improve the accuracy of decisions. These algorithms can reduce the need for human interaction and have a wide range of applications. Furthermore, by identifying specific tendencies, more accurate projections based on historical data can be made. These algorithms can also be used to identify and disrupt the chain of infection transmission [5].

9.3 HAZARD IDENTIFICATION AND DETECTION IN WEB PAGES

Through improvements in information retrieval, aggregation, and discovery, Web 2.0 has completely changed the web and how people use it. Improvements have also been made in several other areas, including web technologies, apps, communication, marketing, and sales, as well as content management and structuring. While these advancements offer many advantages, they also jeopardize basic security and confidentiality requirements. Organizations and laypeople are increasingly facing serious security risks due to the vulnerabilities and security risks brought about by these enhancements [6].

Phishing is becoming increasingly common because it's so easy to deploy. Hackers only need to copy a genuine website and deliver it to their victims via email, using social engineering techniques to entice them in and win their trust. Hackers take advantage of ordinary people's ignorance about internet browsing and the function of a web page's universal resource location (URL). This feature makes it possible for hackers to craft harmful URLs, such as ones that are extremely lengthy or contain questionable characters [7]. One kind of fraud to obtain user credentials is phishing. For financial gain, the attackers obtain sensitive and private user data. Phishing is a widespread issue that impacts various industries, including digital marketing, e-commerce, banking, and online business. It is typically executed using cloned websites and spam emails. The phishers steal users' personal data while they browse the targeted website [8].

A study compares the effectiveness of stacking models and supervised learning techniques for identifying phishing websites. This study's primary contribution is to increase classification accuracy by using PCA-proposed features and stacking the best classifiers. With the suggested features N1 and N2, stacking random forest and Neural Network bagging performed better than all other classifiers, achieving 97.4% accuracy [8].

The data set of phishing websites is used for the research. 11,055 site hits and 32 pre-processed attributes make up the data collection. Usually, artificially taught models are used to extract these features. The validity and consistency of the extracted traits are critical to the investigation. To assess the suggested method's applicability in a real-time situation, it might be integrated with different feature extraction methods in the future [8].

By examining the provided URLs, the author offers a deep reinforcement learning-based model for identifying phishing websites. The model is capable of automatically adjusting to modifications in the URL structure. It presents an automated method for URL-based phishing detection based on reinforcement learning. In order to make the system more dynamic, this deep learning version of the RL algorithm is a complementary strategy to the current phishing detection approaches [9].

9.4 UNDERSTANDING RISKS IN ONLINE PAYMENTS

Frauds involving credit cards are simple and easy to target. Online payment options have expanded due to e-commerce and numerous other websites, raising the possibility of online fraud. Due to an increase in fraud, academics have begun to examine and detect fraud in online transactions using a variety of machine learning techniques [10]. Most people around the globe are aware of frauds because they have been in the news for a few years due to various issues, primarily credit card fraud. The credit card dataset exhibits a significant imbalance as a result of the higher proportion of valid transactions relative to fraudulent ones. While certain on-card payments are safer due to banks switching to EMV cards, which are smart cards that store data on integrated circuits rather than magnetic stripes, card-not-present fraud rates have not decreased [10].

The researcher proposed a novel approach to fraud detection in which each cardholder's profile is created by extracting behavioral patterns from consumer groups based on transactions. Subsequently, distinct classifiers are implemented on three distinct groups, and rating scores are produced for each classifier type. The system can promptly adjust to new cardholder transaction behaviors as a result of these dynamic parameter modifications. Afterwards, a feedback system to address concept drift will be implemented. It has been found that the Matthews Correlation Coefficient was the most effective metric for handling dataset imbalance. MCC wasn't the only option available. The classifiers were discovered to be operating more effectively than previously by using SMOTE to balance the dataset. Using one-class classifiers, such as one-class SVM, is an additional method of managing imbalanced datasets. The last finding is that the algorithms that produced superior outcomes were random forest, decision tree, and logistic regression [10].

9.5 CYBER THREATS ONLINE

Social Engineering: Due to its reliance on human error rather than technological flaws, social engineering is still one of the riskiest hacking methods used by cybercriminals. This increases the threat of these attacks because it is far simpler to fool a person than to compromise a security system. And it's obvious that hackers are aware of this—85% of data breaches involve human involvement, according to Verizon's Data Breach Investigations report [11].

Social engineering techniques will be a crucial tool for gaining access to employee credentials and data in 2023. Emails are the first step in more than 75% of targeted intrusions. After phishing, ransomware and the use of compromised credentials are the next two most common reasons for data breaches. Email impersonation and phishing are always changing to include new strategies, technologies, and trends. For instance, assaults related to cryptocurrencies increased by almost 200% between October 2020 and April 2021, and they will probably continue to pose a serious risk as the value and appeal of Bitcoin and other blockchain-based currencies rise [11].

Third-Party Exposure: Cybercriminals can circumvent security systems by breaking into networks that are less secure and belong to outside parties that have special access to their main target. A significant instance of a third-party breach happened at the start of 2021 when hackers exposed personal information from more than 214 million accounts on Facebook, Instagram, and LinkedIn. By hacking Socialarks, a third-party contractor with privileged access to all three firms' networks, the hackers were able to obtain the data [11].

As businesses increasingly rely on independent contractors to finish tasks that were formerly performed by full-time workers, third-party intrusions will pose an even greater threat in 2023. The ability to access networks will remain a priority for criminal groups. In April 2021, hackers gained access to the U.S. Colonial Pipeline with hacked credentials and a VPN that did not use multi-factor authentication. As a result, they had to pay $5 million in Bitcoin to get access back [11]. Over 50% of organizations are more ready to hire freelancers because of the transition to remote work brought on by COVID-19, according to a 2021 labor trends analysis. Security issues posed by a distributed or remote workforce will persist for both large and small organizations [11]. According to FBI reports, there has been a 300% surge in cyberattacks after COVID-19. According to the report, 53% of respondents concur that working remotely has made it simpler for hackers and other cybercriminals to exploit people. According to a cybersecurity company called CyberArk, 96% of businesses provide these outside companies access to vital networks, giving hackers potentially unprotected access to their data [11].

Configuration mistakes: It's highly likely that even professional security systems have one or more software installation and setup errors. Rapid7, a cybersecurity software company, conducted 268 trials and discovered exploitable misconfigurations in 80 percent of them through external penetration testing. In tests where the attacker had inside system access (trials imitating access via a third party or penetration of a physical office), the percentage of exploitable configuration errors jumped to 96%. [11]. In 2023, the COVID-19 pandemic's combined effects with sociopolitical unrest and persistent financial strain are anticipated to cause more employees to make careless blunders at work, which will provide cybercriminals additional possibilities to exploit. As a result of the epidemic, 81% of workers have reported mental health problems, and 65% of workers claim that their mental health has negatively impacted their ability to do their jobs, according to a Lyra Health report. This pressure will only make the problem worse: according to the Ponemon Institute, half of IT professionals acknowledge they have no idea how well the cybersecurity tools they have installed truly function, which implies that at least half of them aren't currently conducting routine internal testing and maintenance [11].

Inadequate Cyber Hygiene: The term "cyber hygiene" describes customs and behaviors pertaining to the use of technology, such as staying away from unsecured WiFi networks and utilizing security measures like multi-factor authentication or a VPN. Regretfully, studies reveal that Americans' online safety practices might use some improvement [11]. About 60% of organizations save passwords in their human memory, while 42% store passwords on sticky notes. For work accounts, more than half of IT professionals (54%) do not demand two-factor authentication, and only 37% of people utilize two-factor authentication for personal accounts. After a data breach, less than half of Americans (45%) say they would change their password, and just 34% say they do it on a regular basis [11].

Due to a rise in remote work, employees are using personal devices that are far more likely to be lost or stolen, sticky note passwords are finding their way into public coffee shops, and systems secured by weak passwords are now accessible from unprotected home networks. Businesses and people who don't enhance their cybersecurity procedures now face far higher risks than they did previously [11]. It's surprising to see that IT pros frequently practice even worse cyber hygiene than the general public: Only 39% of people in general report they reuse passwords across accounts at work, compared to 50% of IT workers [11].

Vulnerabilities in Clouds: Although one may assume that as cloud security would improve as time goes on, IBM reveals that over the past five years, cloud vulnerabilities have increased by 150%. More than 90% of the 29,000 breaches examined in the study, according to Verizon's DBIR, were brought on by web app breaches. With a 41% increase from $595 million in 2020 to $841 million in 2021, cloud security is now the cybersecurity market area with the strongest rate of growth, according to Gartner [11].

Although experts initially projected a widespread return to the workplace, increases in newly discovered COVID variants and breakthrough case rates have made this scenario less likely [11]. As a result, there is little chance that the growing threat of cloud security breaches will abate by 2023.

The use of "Zero Trust" cloud security architecture is one of the most recent advancements in cloud security. Zero Trust systems are made to operate as if the network has already been breached; rather than allowing consistent access to known devices or devices inside the network perimeter, they enforce necessary verifications at each stage and with each sign-in. In 2021, this security approach became well-liked, and in the upcoming year, it is probably going to be widely used [11].

Mobile Device Vulnerabilities: An increase in the use of mobile devices was another trend brought on by the COVID-19 pandemic. In addition to the fact that remote workers use their mobile devices more frequently, pandemic specialists have urged widespread usage of mobile wallets and contactless payment methods to reduce the spread of germs. Cybercriminals have a greater target when there are more users [11]. The rise in remote work has increased the vulnerability of mobile devices, prompting more businesses to adopt bring-your-own-device rules. A security event involving a malicious mobile application downloaded by an employee occurred in 46% of firms during 2021, according to Check Point Software's Mobile Security Report [11].

Mobile Device Management technologies, which are ironically intended to help businesses control mobile devices in a way that protects corporate data, have also become a target for cybercriminals. Because MDMs are linked to the company's whole mobile device network, hackers can utilize them to attack every employee at once [11].

Internet of Things: With 70% of households owning at least one smart device, the epidemic forced over 25% of American workers to relocate their job from the office into their homes. As expected, this led to an increase in attacks against smart, or "Internet of Things (IoT)" devices; between January and June of 2021, approximately 1.5 billion breaches occurred [11].

The lack of good cyber hygiene practices among the typical American coupled with IoT connectivity creates a world of vulnerability for hackers. Experts predict that a smart house with a variety of IoT gadgets may be the subject of up to 12,000 hacking attempts in a single week. The average smart device is attacked within five minutes of connecting to the internet [11].

According to research, between 2021 and 2025, the number of smart gadgets ordered will double, expanding the network of access points that can be used to compromise corporate and personal networks. Around 3.5 billion cellular IoT connections are anticipated by 2023, and by 2025, experts project that IoT-based cyberattacks will account for more than 25% of all enterprise cyberattacks.

Ransomware: Although ransomware assaults are not new, they have been much more costly recently. The average ransom fee increased from $5,000 to $200,000 between 2018 and 2020. Businesses incur additional costs as a result of ransomware attacks, such as lost revenue as hackers extort users for ransom (system outages following a ransomware attack typically last 21 days) [11].

Sixty-six percent of cybersecurity professionals surveyed in 2021 stated that a ransomware attack caused their companies to lose a sizable amount of cash. After a ransomware assault, 29% of respondents indicated their companies were obliged to remove employees, and one in three reported their organization lost top leadership due to resignation or dismissal [11].

Attacks using ransomware will continue and change as criminal groups try to get around the OFAC blacklist and use coercive methods to get money. Actually, attackers may already sign up for services called "Ransomware-as-a-Service," which let users utilize pre-made ransomware tools to carry out attacks in return for a cut of all successful ransom payments [11].

Cybercriminal organizations, like respectable software businesses, are always adding tools to their arsenal for both themselves and their clients. One such tool is designed to facilitate and expedite the process of exfiltrating data. Threat actors also occasionally use the technique of rebranding their ransomware, altering minor components of it in the process. Microsoft reports that less than four hours are needed for 96.88% of ransomware attacks to successfully infiltrate their target. Malicious software with the fastest speed can take over a company's system in less than 45 minutes [11].

Important Statistics on Ransomware: In 2021, ransomware cost the global economy \$20 billion. By 2031, that amount is anticipated to increase to \$265 billion. Ransomware affected 37% of all enterprises and organizations in 2021. An average of \$1.85 million was spent by firms in 2021 recovering from a ransomware attack. Only 65% of ransomware victims receive their data recovered after paying the ransom, making up 32% of all victims [11].

Just 57% of companies are able to successfully restore their data from a backup.

9.6 CYBER ATTACKS IN SOCIAL MEDIA

Social networks have developed into an online gathering spot for the public in recent years. Regrettably, phishing assaults occur when users interact via social networks [8]. Most businesses use event management and security information systems to keep an eye on their infrastructures and be prepared for cyberattacks. These systems rely on cyber-threat intelligence streams to supply them with up-to-date and pertinent information on threats, updates, and patches. Social media sites like Twitter/X are examples of open-source intelligence platforms that may compile a large number of cybersecurity-related sources. We need scalable and effective systems that can find and summarize pertinent information for specific assets in order to process such information streams [12].

Deep neural network designs were suggested by the study team to carry out the essential functions of a processing pipeline that gathers timely, pertinent, and focused security-related data from Twitter/X. The recommended system can collect tweets from a range of accounts, filter them using a list of keywords that define the infrastructure it will watch, choose the tweets that include pertinent information, and find important information within these tweets [12].

Convolutional and bidirectional long short-term memory neural networks are implemented in the model. It confirms that deep neural network architectures perform better than established approaches by comparing the performance of the proposed methodology to those methodologies. To validate the approach, three case studies provided by two global and one national private organizations were employed. The named entity recognition BiLSTM model achieved an average F1-score of 92% in recognizing specified labels, while the convolutional neural network binary classifier scored an average TPR and TNR of 92% throughout the three case studies [12].

The important study [13] assesses how sentiment variables affect social media bot identification algorithms, as well as how they could affect confirmation bias and the backfire effect. Model accuracy in identifying social bots is enhanced by a new collection of sentiment variables derived from postings made by online users. Additionally, using a new set of attributes yields outstanding results for Dutch tweets in addition to English tweets. For Twitter bot detection, the suggested strategy considers both an account-level approach (using a fresh set of emotion features) and a group-level approach (using existing network features).

9.7 CHALLENGES AND SOLUTIONS

In order to fully utilize technology for risk management, organizations may want to think about using the following strategies: [3].

Platforms for Integrated Risk Management: Adopt integrated risk management systems with the capacity to centralize risk data, facilitate thorough risk assessments, and offer real-time risk monitoring. These platforms enable an organization to handle risks proactively and with a holistic perspective.

Advanced Analytics and AI: To find hidden trends, anticipate future hazards, and aid in decision-making, apply advanced analytics approaches like predictive analytics and machine learning. Risk management solutions with AI capabilities can automate risk assessments and offer insightful data.

Cybersecurity Solutions: To guard against online risks and guarantee data privacy, invest in strong cybersecurity solutions. To protect important assets, use technologies such as vulnerability screening tools, endpoint protection, and intrusion detection systems.

Tools for Risk Visualization and Reporting: Make use of data visualization tools to convey risk information in an understandable and straightforward way. Stakeholders can comprehend risks, keep an eye on important risk indicators, and make data-driven decisions with the help of dashboards and reporting tools.

Continuous Monitoring and Alerts: Put in place automatic monitoring systems that help with real-time response, alerts for anomalies, and ongoing risk assessment. This makes it possible for organizations to minimize any effects and act immediately [3].

9.7.1 Unlocking the Power of Technology in Risk Management

Organizations can achieve incredible advantages by integrating technology into their risk management procedures:

1. Better Risk Insights: Data-driven strategies and advanced analytics offer more in-depth understanding of risks, empowering businesses to take preventive measures and make wise choices.
2. Rapid Risk Response: To minimize possible harm and guarantee company continuity, real-time monitoring, alarms, and automated workflows enable quick risk response.
3. Enhanced Efficiency: Tasks related to risk management can be automated to cut down on manual labor, simplify procedures, and raise overall operational effectiveness.
4. Making strategic decisions: Organizations are empowered to make strategic decisions based on precise and current risk information thanks to technology-enabled risk management.
5. Resilience and Business Continuity: Organizations can improve their resilience, adjust to new hazards, and maintain business continuity during difficult times by utilizing technology [3].

9.8 RISK PREDICTIONS IN FUTURE

When speculating about the future of cybersecurity, one important thing to remember is that anything can happen at any time. The industry changes every year. As cyber dangers develop, so do the defenses against them; these tools are always improving to better protect ever-more-complex networks [14].

The cybersecurity business, even for the bad guys, has seen significant transformation in the previous several years. The swift digital revolution resulting from evolving office settings has expanded the window of opportunity for attackers to select targets. Attackers can now purchase or rent the tools they need for an assault, thanks to the expanding cybercrime-as-a-service (CaaS) market. Time can now be better spent researching and strategically focusing on businesses that are more likely to pay a ransom or otherwise offer a higher return on investment [14].

With laws like the California Consumer Privacy Act (CCPA), the General Data Privacy Regulation (GDPR), and the Personal Information Protection and Electronic Documents Act (PIPEDA) now in effect, victims of data breaches risk fines in the event that private information is revealed. Attackers are taking advantage of this and changing their ransom demands to make it more desirable to pay up than to pay the fine imposed by the law [14].

Misuse of legal and open-source tools: Dual-use tools are regularly developed and maintained by legal penetration testing communities, which makes them useful for a range of sophisticated assaults that would take years to create and test. Off-the-shelf technologies are frequently more affordable and easier to conceal in the commotion of network activity, as demonstrated by a number of recent releases of significant viruses that required years and millions of dollars to build [14].

New and developing cybersecurity trends: Regardless of size or sector, organizations should continue to be aware of the increasing number of trends and possible risks [14].

1. The first is cybercrime-as-a-service, or CaaS. A lone attacker has access to the tools and expertise amassed by thousands, if not millions, of hackers and cybercriminals thanks to the cybercrime-as-a-service economy. This makes it simple for novice hackers to launch sophisticated assaults quickly. Because malicious actors modify their strategies to remain undetected, CaaS markets persist in their operations even after being taken down by law enforcement on multiple occasions [14].

2. Malware automation: As a result of current trends compelling the cybersecurity sector to catch up, malware attacks are becoming more automated. Security professionals no longer have to cope with lone hackers trying their mettle with difficult-to-perform attacks. Hackers can now automate thousands of attacks every day by using a machine to carry out cybercrime activities. Because ransomware is so widespread, media attention appears to be limited to the biggest incidents [14].

3. Polymorphic malware: An increasing amount of malware variants now exhibit polymorphic traits, which means they modify their distinguishing characteristics on a regular basis to better evade detection by security teams and standard procedures. Numerous cloud-based application services use mutable code to enable them to remain hidden [14].

4. Risks and dangers posed by third parties: As businesses continue to expand their reach and integrate digital technologies, many outsource parts of their IT and security support requirements to outside parties. As previously noted, depending on outside parties raises cybersecurity risks, particularly for businesses without a risk management strategy in place [14].

Long-term worries about cybersecurity: A few risks and developments that could shape cybersecurity after 2023 include the following: [14]

1. An increase in the usage of the internet of things (IoT): As more individuals incorporate IoT technology into their daily lives, its usage will rise over the next five years. IoT analytics research indicates that there were 10 billion connected devices in 2019; by 2025, that number might triple to 30.9 billion. For further background, it should be noted that IoT connections surpassed non-IoT connections in 2019. IoT devices continue to have relatively lax security safeguards even though they link to networks and other devices that access extremely sensitive data. The extra security precautions required to keep these devices—as well as everything they're connected to—secure are already difficult for many firms to offer [14].

2. Pay attention to social engineering strategies: In five years, internet communications should be more secure, especially with the possible emergence of quantum networks that will reduce the significance of network-based threats. The persistent problem of human error is one. Whether on purpose or not, workers will continue to facilitate data loss, and hackers will continue to use social engineering techniques like phishing and business email compromise [14].

3. The evolving nature of financial fraud: As a result of payment modernization, financial transactions may soon be conducted nearly exclusively online, needing the assistance of numerous platforms and techniques. Regulations will take time to catch up with these platforms, which are probably going to be less centralized. Financial institutions will face a wider range of threats as a result, leading to an increase in fraud-related security solutions that concentrate on blockchain technology, digital currencies, and real-time payment security [14].

4. Difficulty prosecuting cybercrime: Although an increasing number of nations are placing a high priority on cybersecurity, law enforcement will find it challenging to bring cybercriminals to justice due to a lack of substantiable evidence for crimes committed online. It will also be more difficult to proactively identify cyber risks due to a shortage of cybersecurity experts [14].

9.8.1 Conclusion

Technology also makes it possible to monitor the risk landscape in real time. Organizations may minimize interruptions and capitalize on opportunities amid uncertainty by anticipating new risks and acting quickly with the support of automated alerts and advanced monitoring systems. Technology has also revolutionized risk assessment by providing a comprehensive perspective that takes into account external and internal elements that affect the risk profile of an organization. With this thorough knowledge, companies may create risk mitigation plans that are more successful and promote long-term resilience. To sum up, companies that incorporate technology innovations into their risk management strategies set themselves up for long-term success by improving flexibility and protecting their resources. To succeed in a changing company environment, it is strategically necessary to combine technology with risk management.

REFERENCES

1. www.ncbi.nlm.nih.gov/books/NBK217572/
2. www.sciencedirect.com/topics/earth-and-planetary-sciences/technological-hazard
3. www.linkedin.com/pulse/role-technology-risk-management-heba-al-haddad/
4. Linardos, V., Drakaki, M., Tzionas, P., & Karnavas, Y.L. Machine Learning in Disaster Management: Recent Developments in Methods and Applications. Machine Learning and Knowledge Extraction, 4, pp. 446–473, 2022. doi: 10.3390/make4020020
5. Chamola, V., Hassija, V., Gupta, S., Goyal, A., Guizani, M., & Sikdar, B. Disaster and Pandemic Management Using Machine Learning: A Survey. IEEE Internet of Things Journal, 8, no. 21, pp. 16047–16071, 2021. doi: 10.1109/JIOT.2020.3044966.
6. Saqib, N. A., Salam, A. A., Atta-Ur-Rahman, & Dash, S. Reviewing Risks and Vulnerabilities in Web 2.0 For Matching Security Considerations in Web 3.0. Journal of Discrete Mathematical Sciences and Cryptography, 24, no. 3, pp. 809–825, 2021. doi: 10.1080/09720529.2020.1857903
7. Mourtaji, Y., Bouhorma, M., & Alghazzawi, D. New Hybrid Framework to Detect Phishing Web Pages, Based on Rules and Variant Selection of Features. International Journal of Internet Technology and Secured Transactions, 10, no. 6, pp. 740–757, 2020.
8. Zamir, A., Khan, H. U., Iqbal, T., Yousaf, N., Aslam, F., Anjum, A., & Hamdani, M. Phishing Web Site Detection Using Diverse Machine Learning Algorithms. The Electronic Library, 38, no. 1, pp. 65–80, 2020. doi:10.1108/el-05-2019-0118
9. Chatterjee, M., & Namin, A.-S. Detecting Phishing Websites through Deep Reinforcement Learning. 2019 IEEE 43rd Annual Computer Software and Applications Conference (COMPSAC), 2019.
10. Dornadula, V. N., & Geetha, S. Credit Card Fraud Detection Using Machine Learning Algorithms. Procedia Computer Science, 165, pp. 631–641, 2019.
11. www.embroker.com/blog/top-cybersecurity-threats/#:~:text=Social%20engineering%20remains%20one%20of,to%20breach%20a%20security%20system.
12. Dionisio, N., Alves, F., Ferreira, P. M., & Bessani, A. Cyberthreat Detection from Twitter using Deep Neural Networks. 2019 International Joint Conference on Neural Networks (IJCNN), 2019. doi: 10.1109/ijcnn.2019.8852475
13. Heidari, M., Jones, J. H. J., & Uzuner, O. An Empirical Study of Machine learning Algorithms for Social Media Bot Detection. 2021 IEEE International IOT, Electronics and Mechatronics Conference (IEMTRONICS), 2021. doi: 10.1109/iemtronics52119.2021
14. https://fieldeffect.com/blog/what-is-the-future-of-cyber-security#:~:text=The%20International%20Data%20Corporation%20(IDC,fuelled%20the%20cyber%20insurance%20market.

10 Practical and Innovative Applications of IoT and IoT Networks (Smart Cities, Smart Mobility, Smart Home, Smart Health, Smart Grid, etc.)

Shaik Salma, Asiya Begum, and Hussain Syed

10.1 WHAT DOES "INTERNET OF THINGS" MEAN?

The "Internet of Things" (IoT) refers to a system of interconnected items such as processors, mobile phones, automobiles, and other commonplace objects that are outfitted with sensors, processing power, and network access. Everything from "smart home" devices like smart thermostats to "smart wearables" like smartwatches and RFID-enabled garments to intricate pieces of industrial machinery and transportation networks can be deemed "smart." In fact, some experts are contemplating constructing entire "smart cities" on top of existing IoT networks.

All of these cutting-edge gadgets may be able to communicate with gateways and other internet-connected devices, as well as with one another, thanks to the Internet of Things, potentially creating a vast, self-sufficient network. This can include managing equipment and operations in businesses, keeping an eye on the environment in farms and other places, and tracking products and shipments in warehouses.

IoT is already impacting a wide range of industries, including manufacturing, transportation, healthcare, and agriculture. As the number of internet-connected devices keeps increasing, the Internet of Things is predicted to play a bigger role in changing our world and how we live, work, and interact with each other.

Temperature, humidity, air quality, energy usage, and machine performance are just a few of the characteristics that may be tracked with the use of IoT devices in an industrial setting. Businesses may enhance their processes and boost their bottom line by analyzing this data in real time to spot patterns, trends, and anomalies.

DOI: 10.1201/9781003452393-10

10.2   WHY IS IOT IMPORTANT?

10.2.1   ENHANCED EFFECTIVENESS

Companies have the opportunity to enhance their operational efficiency and productivity through the implementation of Internet of Things devices for process automation and optimization. Utilizing IoT sensors, for instance, enables real-time monitoring of equipment health, which can lead to increased uptime and reduced maintenance expenses.

10.2.2   DECISION-MAKING BASED ON DATA

The massive volumes of data generated by Internet of Things devices can serve as valuable guides for numerous business decisions and even the creation of entirely novel business models. By scrutinizing this data, businesses can enhance their strategies, refine product development, and optimize resource allocation. This analysis provides insights into customer behaviour, market dynamics, and operational efficiency, driving informed decision-making.

10.2.3   COST-SAVINGS

The Internet of Things can help organizations cut expenses and boost profits by eliminating redundant manual operations. To cut expenses and increase sustainability, IoT devices can be used to track and adjust energy consumption.

10.2.4   BETTER INTERACTIONS WITH CUSTOMERS

By using information collected from IoT devices, companies may better understand their consumers' habits and tailor their offerings to them. Using IoT sensors, shops can monitor customers' foot traffic and tailor promotions to their preferences.

10.3   THE INTERNET OF THINGS (IOT)

The IoT can create significant human value by enhancing convenience, safety, sustainability, and efficiency in several spheres of our lives. Here's how IoT connectivity can translate into human value:

a) **Convenience and Efficiency:**

Smart Homes: IoT devices in homes can automate routine tasks, such as adjusting the thermostat, turning off lights, and even reordering household supplies when they run low. This automation reduces the burden of mundane chores, allowing people to focus on more meaningful activities.

Smart Assistants: Voice-controlled IoT devices like Amazon Echo and Google Home make it easier to access information, control smart devices, and perform tasks like setting reminders and making shopping lists.

b) **Well-being and Health:**

Wearable Health Devices: IoT wearables like fitness trackers and smartwatches can monitor vital health metrics, track fitness goals, and provide early warnings for health issues, promoting healthier lifestyles and timely medical interventions.

Remote Patient Monitoring: IoT enables doctors to remotely monitor their patients' health, decreasing the need for frequent hospital visits and enhancing the quality of care for people suffering from chronic illnesses.

c) **Safety and Security:**

Smart Security Systems: IoT-connected security cameras, doorbell cameras, and motion detectors provide real-time monitoring and alerts, enhancing home and business security.

Emergency Response: IoT can be integrated with emergency services to provide faster response times during accidents or medical emergencies.

d) **Energy Efficiency and Sustainability:**

Smart Energy Management: IoT enables better control of energy consumption by optimizing heating, cooling, and lighting systems. This not only reduces energy bills but also contributes to a more sustainable environment by conserving resources.

Environmental Monitoring: IoT sensors can help track air and water quality, detect pollution, and monitor climate changes, leading to improved environmental stewardship.

e) **Transportation and Mobility:**

Connected Vehicles: IoT connectivity in cars can improve traffic management, reduce accidents through real-time safety alerts, and enable autonomous driving, making transportation safer and more efficient.

Public Transportation: IoT has the potential to improve public transportation systems by increasing their dependability and usability, which would ultimately reduce air pollution and traffic congestion.

f) **Accessibility and Inclusivity:**

IoT technology can assist individuals with disabilities by enabling voice-controlled devices, smart home automation, and wearable devices that enhance accessibility and independence.

g) **Business and Industry:**

Supply Chain Optimization: IoT helps businesses improve supply chain efficiency, reduce waste, and ensure the timely delivery of goods and services.

Employee Productivity: IoT applications in the workplace can enhance employee productivity through smart office environments and process automation.

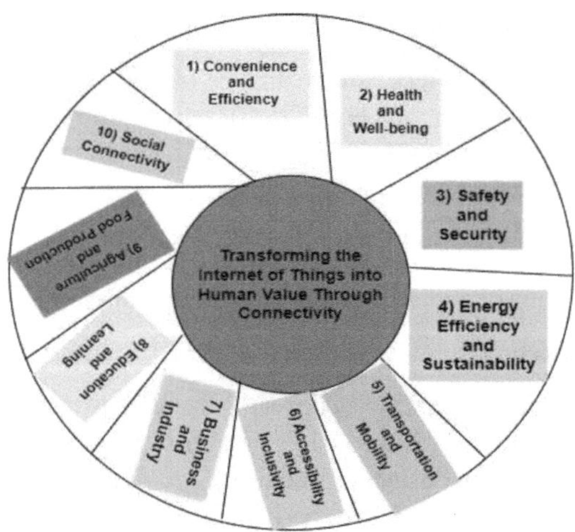

FIGURE 10.1   Transforming the internet of things into human value through connectivity.

h) Education and Learning:

IoT can transform education by providing personalized learning experiences through connected devices and tools, making education more engaging and effective.

i) Agriculture and Food Production:

IoT in agriculture optimizes resource usage, increases crop yields, and ensures food safety, contributing to food security and sustainable farming practices.

j) Social Connectivity:

IoT-connected social platforms and devices enable people to stay connected with friends and family, fostering social well-being and reducing feelings of isolation.

In summary, the IoT has the potential to greatly improve the quality of life by enhancing convenience, safety, health, and sustainability across various aspects of society. Its ability to collect and analyze data in real time allows for more informed decision-making, leading to better outcomes for individuals and communities.

10.4   WHAT IS THE HISTORY OF IOT?

The IoT has a rich history spanning several decades, marked by various phases of technological advancement. Below is an overview of this evolution:

Conceptual Origins (Late 20th Century): The roots of IoT can be traced back to the late 20th century, when the idea of connecting physical objects to the internet began to take shape. Notably, in 1982, a modified

Coca-Cola vending machine at Carnegie Mellon University became one of the earliest internet-connected devices, allowing remote monitoring of its contents.

Early Internet-Connected Devices (1990s): The 1990s witnessed the nascent stages of connecting devices to the internet. Innovations like webcams and early home automation systems started emerging, enabling remote surveillance and control. The term "Internet of Things" was coined by Kevin Ashton in 1999 during his tenure at Procter & Gamble, specifically to describe the concept of Radio-Frequency Identification (RFID) technology.

Proliferation of Wireless Technologies (2000s): The 2000s brought about a proliferation of wireless technologies like Wi-Fi and Bluetooth. These technologies made it easier to connect devices without the constraints of wired connections, fostering the growth of IoT applications, particularly in areas such as home automation and industrial monitoring.

Emergence of IoT Platforms (2010s): The 2010s marked a significant expansion of IoT applications and platforms. Companies like Google, Apple, and Amazon introduced smart home ecosystems, while various industrial sectors adopted IoT for purposes such as asset tracking, predictive maintenance, and supply chain optimization.

Standardization Efforts (2010s): To ensure interoperability and bolster security, various standardization initiatives emerged. Key organizations such as the IETF, the IEEE, and industry consortia like the Industrial Internet Consortium (IIC) actively worked on developing standards and protocols for IoT.

5G and Edge Computing (2020s): The 2020s have seen the deployment of 5G networks and significant advancements in edge computing, further accelerating the growth of IoT. These technologies offer lower latency and enhanced data processing capabilities, making real-time IoT applications increasingly viable.

Privacy and Security Concerns: As IoT adoption has surged, so have concerns regarding privacy and security. The vast amount of data collected and shared by IoT devices has raised important questions surrounding data protection and cybersecurity.

Current and Future Trends: Today, IoT is deeply integrated into various industries, including healthcare, agriculture, transportation, and smart cities. The ongoing evolution of IoT is characterized by developments in artificial intelligence (AI) and machine learning, which enhance its capabilities. Emerging trends include edge computing, IoT analytics, and heightened security measures, all of which are poised to shape the future of IoT.

In summary, the history of IoT illustrates the continuous advancement of technology and connectivity, leading to a world where an ever-expanding array of devices are interconnected, offering enhanced convenience and efficiency across diverse aspects of life and industry.

10.5 THE TOOLS THAT MAKE THE INTERNET OF THINGS POSSIBLE

10.5.1 THERE ARE A LOT OF TECHNOLOGIES THAT WORK TOGETHER TO MAKE IoT POSSIBLE

Sensors and actuators: Environment-changing factors including temperature, humidity, light, motion, and pressure can all be detected by sensors. Actuators are machines that can trigger a certain action, such as the activation of a motor or the opening of a valve. In order for technologies and other devices to interact with the real world, these components are essential to the IoT. When sensors and actuators cooperate to find solutions to problems without human intervention, we have automation.

Connectivity technologies: Connectivity to the internet is required for IoT devices for transmitting data from sensors and actuators to cloud-based platforms. Within the realm of the Internet of Things, various communication technologies are employed, including Wi-Fi, Bluetooth, cellular networks, Zigbee, and LoRaWAN, among others.

Cloud computing: The massive amounts of data produced by Internet of Things devices are sent to the cloud where they may be stored, processed, and analyzed. When it comes to storing and analyzing this data, as well as developing and deploying Internet of Things. Cloud computing platforms and apps provide the necessary infrastructure and capabilities.

Big data analytics: Businesses require sophisticated analytics tools to sift through the mountains of data produced by IoT devices and draw useful conclusions. Some examples of such instruments are data visualization programs, predictive analytics models, and machine learning algorithms.

Security and privacy technologies: IoT security and privacy are becoming more crucial as IoT deployments proliferate. Intrusion detection systems, access controls, and encryption are some of the technologies used to safeguard IoT devices and the data they produce against online attacks.

10.5.2 THE HARDWARE, SENSORS, CONNECTIVITY, AND SOFTWARE THAT MAKE UP THE IoT MAKE IT POSSIBLE TO DO WHAT IT DOES. HERE IS A BREAKDOWN OF HOW THE IoT FUNCTIONS

Sensors and Devices: The IoT starts with physical devices and sensors. Simple sensors (e.g., temperature sensors, motion detectors) to more complicated devices (e.g., smart thermostats, wearable fitness trackers) are examples of these devices. These sensors collect data from the surroundings or the thing being monitored.

Data Collection: IoT sensors are always gathering information from the outside environment. For instance, a fitness tracker tracks your steps and heart rate, while a smart thermostat measures the temperature and humidity in a house.

10.5.3 CONNECTIVITY: THE INFORMATION GATHERED BY IoT DEVICES IS ROUTED TO A CENTRAL HUB OR CLOUD-BASED PLATFORM. THIS CONNECTIVITY CAN BE DONE IN A VARIETY OF WAYS

a) **Wi-Fi:** Many IoT devices connect to the internet via home or business Wi-Fi networks.

b) **Bluetooth:** Certain gadgets link to adjacent smartphones or other devices via Bluetooth.

c) **Cellular Networks:** Cellular networks can be used by devices to send data when they must be mobile or in remote locations.

d) **Low-Power Wide-Area Networks (LPWANs):** These networks are designed for IoT devices that require low-power, long-range communication.

e) **Ethernet:** Wired Ethernet connections are typical for Internet of Things devices in industrial settings.

f) **Data Processing:** The data is processed after it is sent to the cloud platform or central hub. Data analysis, storage, and occasionally machine learning techniques are involved in this. The processing's goal is to draw insightful conclusions from the unprocessed data.

g) **Decision-Making and Control:** Based on the processed data, the IoT system can make decisions automatically, depending on the application. For example, a smart thermostat may change the temperature according to the homeowner's preferences and the weather prediction.

h) **User Interface:** A user interface is a feature of many IoT applications; they are either web dashboards or smartphone apps. Through these interfaces, users can communicate with their IoT devices, view data, and adjust settings from a distance.

i) **Feedback Loop:** IoT devices frequently give consumers feedback or alerts in response to the data they gather. For instance, if a security camera notices movement while the homeowner is gone, it may sound an alert.

j) **Continuous Operation:** IoT devices, as long as they are powered and connected, work nonstop, gathering and sending data. Monitoring and control in real time are made possible by this continuous operation.

k) **Security and Privacy:** Ensuring Internet of Things data security and privacy is essential. Access controls, encryption, and authentication are put in place to safeguard data and stop illegal access.

l) **Scalability:** IoT systems can be made more capable by adding additional devices and sensors to the network as needed because they are frequently scalable.

10.6 IOT EMERGENCE

To put it briefly, the Internet of Things functions by gathering data from sensors and physical devices, sending it to a central platform, processing it to help with decision-making and insights, and enabling interfaces that let users communicate with and manage an IoT apparatus. Optimizing convenience, efficiency, safety, and other facets of human existence is the aim of using data from IoT devices.

The emergence of the IoT is a significant technological development that has transformed various industries and aspects of daily life. Here is an overview of the key factors contributing to the Internet of Things emergence:

Advancements in Connectivity: The IoT has its roots in the widespread use of high-speed internet, wireless networks (such as 4G and 5G), and low-power communication technologies (like Bluetooth and Zigbee). Devices and the internet can communicate seamlessly thanks to these technologies.

Miniaturization of Hardware: It is now affordable to integrate computing power into commonplace things thanks to the shrinking of sensors, CPUs, and other hardware components. These compact, energy-efficient parts are essential to Internet of Things devices.

Cost Reduction: IoT devices are now more reasonably priced and available to both consumers and enterprises thanks to the declining cost of hardware components, especially microcontrollers and sensors.

Data Analytics and Cloud Computing: The infrastructure to switch and retain the large capacities of data produced by Internet of Things devices has been made possible by developments in cloud computing and data analytics. Cloud platforms provide scalable and reasonably priced ways to manage IoT data.

Machine Learning and AI: Intelligent decision-making can be facilitated by combining IoT data with ML and AI algorithms, which can yield important insights. AI improves IoT device operation by increasing predictability and adaptability.

Standardization: The integration of various IoT devices and ecosystems has been made easier by the creation of industry standards and protocols for IoT communication and interoperability.

Security Improvements: The emphasis on security has increased along with the proliferation of IoT. Best practices and standards for IoT security have developed to shield data and devices from online attacks.

Consumer Demand: IoT devices are being used by consumers more frequently due to their efficiency, safety, and convenience. IoT technology innovation and investment have been spurred by this demand.

Industry Applications: Numerous sectors, including industry, healthcare, agriculture, and transportation, have realized how the IoT can boost productivity, cut expenses, and increase safety. This has prompted large investments in IoT solutions designed to meet the demands of industries.

Government Initiatives: Numerous nations have started initiatives and policies to encourage the adoption of IoT, especially in areas like smart cities where IoT may enhance services and infrastructure.

Environmental and Sustainability Goals: IoT is essential for controlling and keeping an eye on environmental resources, advancing sustainability objectives, and tackling climate change issues.

Startups and Innovation Ecosystems: IoT innovation has been greatly aided by the startup ecosystem, which has produced new gadgets, software, and business models.

Global Connectivity: IoT is a worldwide phenomenon since it is not restricted by geographical borders. It makes it possible to remotely monitor and manage assets and devices from anywhere in the globe.

Consumer Education: Customers are now better equipped to make decisions about adopting IoT devices thanks to increased knowledge and education on the advantages and risks of IoT.

The IoT is a technological model shift that offers new prospects for efficiency, automation, and data-driven decision-making across a variety of businesses and daily life. It represents the convergence of the physical and digital worlds. Looking ahead, it is anticipated that IoT will become even more important as technology develops.

10.7   CHARACTERISTICS OF IOT

The IoT is characterized by several key features and attributes that distinguish it from other technological paradigms. These features contain:

Connectivity: IoT devices can communicate and share data with other systems & devices because they are connected to the internet or other networks. Both physical (like Ethernet) and wireless (like Wi-Fi, cellular, and Bluetooth) communication is possible.

Sensing and Data Collection: IoT devices can collect data from the physical environment because they have sensors and data gathering capabilities built into them. Numerous characteristics, including temperature, humidity, motion, and location, can be measured using these sensors.

Data Processing: Microcontrollers and embedded processors are used in Internet of Things (IoT) devices. These devices can process and analyze data locally before sending it to a central system or the cloud. By using local processing, latency may be decreased and bandwidth can be preserved.

Remote Monitoring and Control: Internet of Things enables remote system and device control and monitoring. IoT devices can be accessed and managed by users via web interfaces or smartphone apps from any location with an internet connection.

Interoperability: IoT systems and devices are made to function together without any issues, irrespective of the connection protocol or manufacturer. To achieve this compatibility, interoperability standards and protocols are essential.

Scalability: IoT ecosystems and networks are scalable, allowing for a high volume of users and devices. For applications ranging from smart homes to smart cities, this scalability is crucial.

Real-Time Operation: Many Internet of Things applications require real-time or near-real-time operation, especially in scenarios like industrial automation, healthcare monitoring, and autonomous vehicles.

FIGURE 10.2   Characteristics of internet of things.

Security and Privacy: IoT devices and data are vulnerable to security threats [1–3], so robust security measures are essential. Encryption, authentication, and access controls are used to protect IoT systems and data. Privacy concerns, especially regarding the collection and use of personal data, also play a significant role in IoT development.

Energy Efficiency: Many IoT devices operate on battery power or have limited energy resources. Therefore, they are designed to be energy-efficient, often using low-power components and sleep modes to extend battery life.

Location Awareness: IoT devices often have location awareness through technologies like GPS or triangulation methods. This enables applications like asset tracking, navigation, and location-based services.

Data Analytics and Insights: IoT generates vast amounts of data. Analytics & ML are used to extract actionable insights from this data, enabling data-driven decision-making and predictive maintenance, among other applications.

Diversity of Applications: IoT is incredibly versatile and finds applications in various industries, including healthcare, agriculture, transportation, manufacturing, smart cities, and consumer electronics. Its adaptability to different use cases is a defining characteristic.

Continuous Evolution: IoT is an evolving field with ongoing technological advancements. New sensors, connectivity options, and standards continue to shape the IoT landscape, leading to innovative applications and solutions.

Cost Efficiency: Over time, the cost of IoT hardware has decreased, making it more accessible to businesses and consumers alike. This cost efficiency has contributed to the widespread adoption of IoT.

These characteristics collectively define the nature of IoT and its potential to transform industries, improve efficiency, enhance convenience, and address various societal challenges.

10.8   ADVANTAGES OF IOT

Many businesses and applications can benefit from the many advantages that the Internet of Things offers. The following are some of the main benefits of IoT:

a. Efficiency Improvement:

Operational Efficiency: IoT enables businesses and organizations to streamline operations, automate tasks, and reduce manual processes, leading to increased efficiency and productivity.

Resource Optimization: IoT sensors and data analytics help enhance resource usage, such as energy, water, and raw materials, reducing waste and costs.

Supply Chain Efficiency: Enhanced logistics, fewer errors, and overall supply chain efficiency are achieved by real-time tracking and monitoring of commodities in the supply chain.

b. Cost Savings:

Reduced Maintenance Costs: Predictive maintenance powered by Internet of Things can significantly reduce equipment downtime and maintenance costs by identifying and addressing issues before they lead to failures.

Energy Savings: IoT-based energy management systems help organizations cut energy consumption and costs by optimizing heating, cooling, lighting, and other systems.

Inventory Management: IoT-enabled inventory tracking reduces overstocking and understocking, minimizing carrying costs and lost sales opportunities.

c. Enhanced Safety and Security:

Workplace Safety: IoT sensors and wearables can monitor workplace conditions and employee safety, reducing accidents and ensuring compliance with safety regulations.

Home Security: Smart home security systems with IoT capabilities provide real-time monitoring and alerts to homeowners, enhancing residential security.

d. Improved Decision-Making:

Data-Driven Insights: A multitude of data is produced by IoT, which can be analyzed to provide insightful information. Businesses and organizations are better able to respond to changing circumstances and make well-informed decisions thanks to data-driven decision-making.

Predictive Analytics: IoT allows for predictive analytics, helping organizations anticipate trends and issues and proactively address them.

e. Convenience and Quality of Life:

Smart Homes: IoT devices in homes offer convenience and comfort by automating tasks like controlling lighting, thermostats, and appliances. They also enhance security and entertainment options.

Health and Wellness: IoT health devices and wearables enable individuals to monitor their health and fitness, leading to healthier lifestyles and better overall well-being.

f. Environmental Benefits:

Sustainability: IoT supports sustainability efforts by monitoring and managing environmental resources, reducing energy consumption, and minimizing waste.

Conservation: IoT sensors are used for wildlife conservation and habitat monitoring, contributing to the preservation of endangered species.

g. Healthcare Advancements:

Remote Patient Monitoring: IoT devices enable remote patient monitoring, decreasing the frequency of hospital stays and enhancing the quality of care, particularly for patients with long-term illnesses.

Telemedicine: IoT facilitates telehealth services, providing access to medical expertise for individuals in remote or underserved areas.

h. Smart Cities:

Urban Efficiency: IoT technologies in smart cities improve traffic management, waste management, and public services, making cities more livable and sustainable.

Safety: IoT-enabled surveillance and public safety systems enhance urban safety and emergency response.

i. Competitive Advantage:

Innovation: Early adoption of IoT can provide a competitive edge by offering innovative products, services, and business models.

Customer Engagement: IoT-driven personalization and enhanced customer experiences can build customer loyalty and satisfaction.

j. Accessibility and Inclusivity: IoT solutions can make technology more accessible to individuals with disabilities, improving their quality of life and participation in society.

These benefits demonstrate the transformative power of IoT across various domains, contributing to increased efficiency, sustainability, safety, and quality of life for individuals and organizations alike.

10.9 PRACTICAL AND INNOVATIVE APPLICATIONS OF IOT AND IOT NETWORKS ACROSS VARIOUS DOMAINS, INCLUDING SMART CITIES, SMART MOBILITY, SMART HOME, SMART HEALTH, AND SMART GRID

The Internet of Things is the next phase of industrial development. We can now digitalize once-physical artifacts thanks to technology. The IoT is the fusion of the "real world" with the "digital world," enabling smooth communication between people, machines, and objects. The Internet of Things enables remote activation and control of items using an internet network architecture, creating secure interaction between the physical world and digital processes. Every facet of contemporary life is being revolutionized by the IoT. Now, as opposed to earlier, IoT is a genuine phenomenon. The effects of widespread internet access extend well beyond mobile gadgets. Physical objects may now automatically record, monitor, and exchange data instead of needing a human to do so in the past thanks to the internet.

A list of possible IoT applications has been developed by the IoT European Research Cluster (IERC) using scientific research, publications, and expert comments. The Internet of Things can be used in a wide range of applications, including smart cities, smart industries, smart transit, smart buildings, smart energy, smart manufacturing, smart environment monitoring, smart living, smart health, and smart food and water monitoring. The short life cycle of manufacturing puts pressure on output and industrial automation, and in many industries, short-term marketing is crucial. Future production techniques will place a higher priority on design flexibility and adaptation. This updated list of IoT appliances shows how IoT can be used in a variety of contexts and explains why it is one of the main trends that will likely be popular over the next five years.

Smart industry

Appliances that communicate with other machines: asset management and automated machine diagnostics.

The tank or vessel's liquid level: keeping an eye on the water and fuel levels in wells and storage containers.

Volume calculator for silos: weight of the goods and the measure of emptiness.

Air quality within and outside building constructions and buildings: In order to ensure the security of both employees and goods, hazardous gas and oxygen concentrations must be closely monitored in an industrial processing facility that produces chemicals [4, 5].

Ozone monitoring: throughout the dry meat processing phase in the food sector, ozone is present.

Temperature measurement and monitoring: managing the temperature of critical components within

10.9.1 Smart Cities

Enhancing efficiency can lead to more resource and energy efficiency in cities. This can be achieved by utilizing a range of sensors that are dispersed across the city and serve different purposes. These sensors can be used for a variety of tasks, including optimizing streetlights, managing waste, managing traffic, and creating smart buildings. Many cities throughout the world, like Oslo, Singapore, Geneva, Zurich, and others, are trying to incorporate IoT and become smarter. The Smart Nation Sensor Platform is used by Singapore, which is recognized as the smartest city in the world, as an example of how to develop smart cities. This platform integrates several areas of public safety, transit, streetlights, urban planning, etc. by using sensors in conjunction with IoT.

Smart Traffic Management: Sensors and cameras connected to the Internet of Things track traffic conditions in real time, enabling adaptive signal timing and optimizing traffic patterns to lessen congestion.

Waste Management: Garbage collection routes can be optimized and costs reduced with the help of smart trash cans that use sensors to provide alerts when they are full.

Environmental Monitoring: By monitoring things like noise and air pollution in real time, IoT networks might help cities become more sustainable.

Smart Parking: When Internet of Things (IoT) sensors are installed in parking lots, they can alert cars to open spots and direct them there.

Public Safety: By giving police access to real-time data, IoT-enabled security cameras and gunshot detection devices make communities safer.

City noise mapping: Real-time monitoring of noise levels within the concentrical zone, with bar areas.

Construction health: Monitoring vibration and material conditions in landmark structures including sculptures, bridges, and monuments [6].

Traffic jam: Pathway improvements based on monitoring pedestrian and vehicle traffic.

Safer cities: Control and prevention systems for fires, as well as video surveillance and public address systems.

Intelligent lighting: Raise lighting sources on the side of the road that are both smart and weather-responsive.

The uses for the Internet of Things vary, and some uses may not be suitable for all users. The needs of various customer subsets behind the wheel vary widely. There are three primary types of consumers to consider from an Internet of Things perspective:

- people as individuals; • citizens as groupings (citizens of a country, state, or city); • businesses.

10.9.2 SMART MOBILITY

Autonomous vehicles, also referred to as smart cars, are a product of the IoT. Modern cars come with a lot of interconnected parts that need to communicate with each other. Examples of these parts include speed and brake controls, antennae, navigation sensors, and more. Internet of Things technology is essential for self-driving cars because it allows for smooth communication between moving parts and precision down to the millisecond. Tesla is creating self-driving vehicles, which are becoming quite well-liked. Tesla Motors is at the vanguard of automotive innovation with its vehicles, which use cutting-edge AI and IoT technologies. They're also quite well-liked! With almost 140,000 sold in the US in 2018, the Tesla Model 3 is the most well-liked plug-in electric car in the nation.

Connected Vehicles: Safety is increased and traffic is decreased because to the Internet of Things connectivity in cars allowing for connectivity between vehicles, remote diagnostics, and traffic updates in real time.

Ride-Sharing and Autonomous Vehicles: Ride-sharing apps and autonomous vehicles are made possible by the Internet of Things, which is transforming urban transportation and making automobile ownership less necessary.

Public Transportation: Public transportation can be improved with the use of Internet of Things technologies, which provide real-time updates on schedules, occupancy levels, and maintenance needs.

Shipment Quality: Monitoring for insurance purposes any damage, vibrations, impacts, or openings in containers.

NFC Payment: Merchants can take online credit card and debit card payments by connecting with a merchant bank or acquirer based on the location or time needed to complete the operations of public transit, museums, galleries, and so on.

Object Location: Examine specific items in big places like ports or repositories.

Keep tabs on the movement of your fleet's cars and equipment: Managing the flow of precious materials such as diamonds, pharmaceuticals, and other potentially dangerous goods.

Detection of incompatible storage conditions: Notifying when containers emit easily flammable substances near others containing explosives.

Vehicle management: Car-sharing companies oversee vehicle usage via smartphones equipped with internet connectivity installed in any vehicle.

Automated automobile diagnostics: Gathering data from the CAN Bus to send real-time alerts about potential risks or provide driving tips to vehicle operators.

10.9.3 SMART BUILDINGS

Liquid presence monitoring: Detecting the presence of liquids in data centers, storage facilities, and critical underground structures to prevent structural damage and chemical deterioration.

Perimeter access management: Implementing controlled access to secure areas and identifying unauthorized individuals in restricted zones.

Indoor climate regulation: Monitoring and regulating temperature, lighting, and indoor air quality parameters such as carbon dioxide levels.

Art and culture preservation: Conducting condition monitoring within museums and art galleries to safeguard cultural artifacts.

Intrusion detection system: Identifying openings in doors and windows and breaches to prevent unauthorized entry, particularly in structures with malicious intent.

Irrigation for homes: The use of an intelligent irrigation and monitoring system in private residences.

10.9.4 SMART ENERGY

Photovoltaic system setup: Overseeing and optimizing the performance of solar power installations [7].

Grid enhancement: Monitoring and managing electricity consumption within the power grid.

Wind turbine power generation: By monitoring and evaluating the power output from wind turbines and establishing bidirectional communication, clients can analyze their usage habits with smart meters.

Radiation levels: Collaborative calculation of radiation levels around nuclear power plants to generate alerts for potential leaks.

Flow measurement: Gauging the water pressure propelling water through pipelines within water distribution systems.

10.9.5 Smart Production

Compost Management: Regulating the temperature and humidity of materials like hay, alfalfa, and straw to prevent fungal and microbial contamination during composting.

Automated Manufacturing Oversight: Implementing automation for replenishing supplies and managing product rotation in warehouses and storage areas, following the FIFO (First-IN, First-OUT) principle.

Progeny Supervision: Implementing controlled reproduction strategies in agricultural animals to ensure their well-being and survival [6].

Toxic Gas Monitoring: Conducting investigations to monitor air pollution within agricultural structures and detect hazardous chemicals in stables [4, 5].

Animal Tracking: Utilizing systems for locating and categorizing animals that graze within extensive stables or open pastures.

Telecommuting: By equipping employees with the necessary equipment to work from home, employers can cut expenses, boost output, and create more jobs. They can also eliminate daily office travel, staff housing, and office maintenance and cleaning.

Production-line Monitoring: Production-line information can be changed to suit company needs using cloud-based solutions, RFID, sensors, video surveillance, remote data exchange, & other technologies.

10.9.6 Smart Environment Monitoring

Air Pollution Reduction: Efforts to minimize the emissions of CO_2 from vehicles, factories, and the release of hazardous gases in agricultural regions [4, 8].

Forest Fire Detection: Utilizing preventive fire condition monitoring and flue gas analysis to identify potential fire hazard zones [9].

Avalanche and Landslide Mitigation: Monitoring soil moisture, Earth's density, and vibrations to identify trends that could lead to hazardous avalanche and landslide conditions.

Wildlife Conservation: Tracking and locating wildlife using GSM/GPS modules, then relaying their locations via SMS for conservation efforts.

Early Earthquake Detection: Implementing distributed control systems in earthquake-prone areas to detect seismic activity in advance.

Ocean and Coastal Monitoring: Deploying an array of sensors on satellites, ships, aircraft, and other platforms for purposes such as tracking fishing vessels, ensuring marine security, and identifying potential oil spills.

Weather Station Networks: Establishing networks of weather stations in agricultural areas to study meteorological conditions, enabling forecasts for changes in air patterns, droughts, and ice formation.

10.9.7 SMART LIVING

Monitoring Water and Energy Usage: Tracking water and energy consumption to receive recommendations for reducing resource usage and costs.

Intelligent Shopping System: Identifying ideal shopping locations based on factors such as product expiration dates, consumer preferences, purchasing habits, and allergy considerations.

Remote Device Control: Enabling the remote activation and deactivation of devices to enhance safety and save energy.

Smart Home Appliances: Examples include transparent LCD display refrigerators that show contents, alert about spoiled food, suggest kitchen supplies, and provide access via a smartphone app [10]. Washing machines can be operated remotely and set to run during low-power cost periods. Smart cooking apps allow temperature adjustments and remote monitoring of oven cleaning features.

Weather Monitoring Station: Utilizing meters capable of transmitting data over long distances to provide real-time information on outdoor conditions like humidity, temperature, atmospheric pressure, wind speed, and rainfall.

Safety Compliance Monitoring: Employing baby monitors, optical tools for capturing images, and home security systems to enhance daily safety and peace of mind at home.

Gas Detection System: Connecting local gas meters to the Internet Protocol (IP) network to provide real-time information about gas usage and the condition of gas pipelines. This can potentially lead to reduced labor and repair costs, improved accuracy, more affordable meter readings, and potentially lower gas consumption when monitoring and assessing water quality.

10.9.8 SMART HOME

Smart homes represent one of the most popular and widely recognized applications of the IoT. The concept involves integrating various home appliances and systems, such as locks, lights, air conditioners, and thermostats, into a unified system that can be controlled conveniently from a smartphone. Many people are drawn to these IoT devices because they offer the flexibility to customize and personalize the appearance and functionality of their homes.

The market for IoT gadgets is rapidly expanding, with approximately 127 new internet-connected devices being introduced every second. Some of the most notable and well-known examples of these devices include Google Home, Amazon Echo

Plus, the Philips Hue Lighting System, and numerous others. It's quite likely that you have either heard of or have already integrated some of these devices into your own home. Additionally, there are several other innovative products available for home automation, such as the Nest Smoke Alarm and Thermostat, the Foobot Air Quality Monitor, and the August Smart Lock, to name just a few. These devices provide homeowners with an array of options for enhancing convenience, security, and energy efficiency within their living spaces.

Home Automation: Smart thermostats, lights, and locks are examples of IoT devices that let users control and improve their energy use and safety from afar.

Voice Assistants: Voice-activated assistants like Alexa and Google Assistant can connect to IoT networks and be used with them. This lets you handle smart home features with your voice.

Health and Wellness: IoT-powered health devices keep an eye on vital signs, remind people to take their medications, and allow for remote consultations, all of which are good for everyone's health.

10.9.9   SMART HEALTH

IoT technology has found numerous valuable applications in the healthcare sector. For instance, it enables physicians to remotely monitor their patients using an interconnected network of tools and devices, eliminating the need for constant physical presence [11]. This remote monitoring is especially beneficial for patients with minor health concerns or those suffering from contagious illnesses like COVID-19.

One prominent application of Internet of Things in healthcare is the utilization of robots. Surgical robots, for example, have become increasingly prevalent. These robots assist surgeons in performing procedures more efficiently, accurately, and with greater precision. Additionally, they offer enhanced control during surgeries.

Cleaning robots equipped with high-intensity ultraviolet light have also gained popularity in healthcare settings. These robots effectively and rapidly sanitize surfaces, a capability that is particularly valuable in current times.

Furthermore, nursing robots have emerged as another category of IoT-powered devices. These robots are capable of handling routine and repetitive tasks that nurses typically perform for numerous patients. This automation reduces the workload on healthcare professionals and minimizes the risk to patients.

In summary, IoT applications in healthcare extend to remote patient monitoring, surgical assistance, sanitization, and task automation through robotic technology, ultimately contributing to improved patient care and healthcare system efficiency.

Remote Patient Monitoring: IoT devices provide constant tracking of patients' health, which helps find problems early and lowers the number of hospitals stays.

Medical Wearables: IoT wearables keep track of vital signs, fitness metrics, and sleep habits, giving people the tools they need to live a healthier life.

Telemedicine: IoT networks support telehealth services, which are helpful in rural or underserved areas because they allow for virtual consultations and medical evaluations from a distance.

Activity Tracking for Elderly Individuals: Utilizing a wireless body area network to measure movement, vital signs, visual acuity, and cellular connectivity, gathering, displaying, and storing activity data to support physical activity in older individuals.

Fall Detection System: Implementing a system designed to promote independent living for seniors or those with disabilities by identifying falls and providing necessary assistance.

Pharmacy Refrigeration Monitoring: Continuously monitoring and controlling storage conditions within cold storage units dedicated to storing medicines, vaccines, and organic materials.

Continuous Patient Monitoring: Providing continuous surveillance of patients' health conditions within healthcare facilities such as hospitals and nursing homes.

Athlete Personal Care: Monitoring vital signs in high-performance environments and sports camps, with fitness and health devices available to measure various metrics, including fitness levels, step counts, weight, and blood pressure.

Chronic Illness Management: Offering remote patient monitoring programs that collect comprehensive patient data, resulting in advantages such as reduced healthcare expenses and temporary hospital stays.

Hand Hygiene Oversight: Employing RFID wristbands integrated with Bluetooth LE tags to track and alert individuals during handwashing, while also collecting data for analyzing patient health outcomes among specific healthcare staff [6].

Ultraviolet Light Measurement: Assessing UV sunlight levels to provide warnings about potential exposure during specific periods.

Oral Care Management: Introducing Bluetooth-enabled toothbrushes paired with smartphone apps to evaluate brushing habits and offer data for personal use or sharing with a dentist.

Sleep Monitoring: Implementing IoT devices placed on or near the bed to monitor typical sleep movements, including breathing, heart rate, and significant body movements, with data accessible through a smartphone app.

Smart Water and Food Monitoring: Utilizing IoT technology for monitoring water quality, detecting leaks, controlling supply chains, and enhancing wine quality through vineyard management [12].

Water Leak Detection: Identifying the presence of water in external tanks and monitoring pressure fluctuations in pipelines.

Water Quality Assessment: Analysing water suitability in natural watercourses and oceans for regional wildlife and human consumption.

Flood Monitoring: Observing changes in river, reservoir, and dam flow rates for flood prediction and management.

Supply Chain Management: Monitoring storage conditions and tracing manufacturing processes to enhance supply chain control and tracking.

Water Resource Management: Collecting real-time water condition and usage data by connecting water meters to an IP network, aiding in efficient resource management.

Wine Quality Enhancement: Improving wine quality by monitoring soil moisture and vine trunk width in vineyards to regulate grape sugar levels and ensure vine health.

Golf Course Irrigation: Implementing selective irrigation in arid areas on golf courses to conserve essential water resources.

Greenhouse Microclimate Control: Regulating microclimatic conditions to boost the production and quality of vegetables and fruits in greenhouses.

Monitoring Water Quality in Agricultural Fields: Enhancing the preservation of food products by implementing thorough monitoring, continuous data collection, statistical analysis, and efficient crop field management practices. This encompasses precise control over factors such as fertilization, irrigation, and electricity consumption.

10.9.10 SMART GRID

Energy Management: Smart meters and grid systems that are connected to the IoT make it easier to distribute energy, cut down on power outages, and encourage the use of renewable energy sources.

Demand Response: During times of high demand, IoT networks let utilities talk to and control connected products, which lowers energy use and costs.

Grid Security: IoT devices and analytics make the power grid safer by finding problems and possible cyber threats.

10.9.11 SMART RETAIL

Retailers may use cutting-edge technology, particularly the Internet of Things (IoT), to give customers a better shopping experience. Retail businesses may improve the shopping experiences of their employees and customers in a number of ways by utilizing IoT. It can cut down on theft, improve in-store interactions, optimize inventory management, and do away with long lines at the checkout. The Amazon Go stores, which provide an IoT-enabled shopping experience, are a great illustration of this. Customers may choose their purchases and leave the store without having to wait in line at the cash register thanks to these retailers' use of IoT to monitor each product. After the customer leaves the business, the entire bill is then automatically paid to their card that is linked to Amazon.

10.9.12 SMART POLLUTION CONTROL

One of the main issues facing most cities worldwide is pollution. It is unclear at times whether we are breathing in smoke or oxygen! IoT can significantly aid in such a scenario; efforts aimed at reducing pollution levels to a more tolerable and sustainable range have become crucial.

This can be accomplished by deploying an array of sensors in conjunction with the IoT to collect data on urban pollutants, including vehicle emissions, pollen levels, wind patterns, weather conditions, traffic volumes, and more. Leveraging this data, machine learning algorithms can generate pollution forecasts for different areas within the city, providing advance notifications to local authorities about potential problem areas. Subsequently, these authorities can take proactive measures to mitigate pollution levels and create a safer environment. IBM's Green Horizons initiative from its China Research Lab serves as an exemplary illustration of this approach.

10.9.13 SMART AGRICULTURE

Food is an essential element of human life, and its absence can lead to dire consequences. Unfortunately, while people in less developed nations like Chad and Sudan suffer from hunger, a significant amount of food goes to waste in affluent countries such as the United States. Leveraging IoT technology is one approach to address this issue and ensure food security for all.

This can be accomplished by initially gathering farm-specific data from a variety of sources, including farm sensors, satellites, and local weather stations. This data encompasses critical factors like soil quality, sunlight levels, seed varieties, and rainfall patterns. Subsequently, by employing machine learning and IoT technology, personalized recommendations can be generated for each farm. These recommendations optimize planting methods, irrigation schedules, fertilizer application, and other variables, ultimately leading to increased crop yields and a significant reduction in global hunger.

SunCulture, a participant in the Microsoft AI for Earth program, exemplifies this efficient approach to improving agricultural operations and addressing food security challenges.

These useful and creative uses of IoT and IoT networks are revolutionizing energy management, houses, transportation, urban life, and healthcare. They raise people's and communities' general standards of living while increasing effectiveness, convenience, and sustainability.

10.10 RISKS AND CHALLENGES IN IOT

IoT has a lot to offer, but there are hazards and difficulties as well. The following are some of the more noteworthy ones:

Risks to privacy and security: Security and privacy are becoming more crucial as IoT devices proliferate. Hackers and other cyberthreats can undermine the security and privacy of sensitive data by targeting many IoT devices. Large volumes of personal data can be collected by Internet of Things devices, which raises privacy and data protection issues.

Interoperability problems: It can be challenging for IoT devices made by various manufacturers to communicate with one another since they frequently employ multiple standards and protocols. This may result in data silos that are challenging to combine and analyze, as well as interoperability problems.

Data overload: Businesses unprepared to handle the huge amounts of data generated by Internet of Things devices may find themselves inundated. It can be quite difficult to analyze this data and derive useful insights, particularly for companies without the required analytics knowledge and resources.

Cost and complexity: Setting up an Internet of Things system can be expensive and complicated, involving large infrastructure, software, and hardware investments. It might be difficult to manage and maintain an IoT system, as it requires specific knowledge and abilities.

Legal and regulatory issues: As IoT devices proliferate, new legal and regulatory issues are arising. Companies must abide by several cybersecurity, privacy, and data protection laws, which differ from nation to nation.

10.11   THE FUTURE OF IOT

The IoT has a bright future ahead of it, with lots of exciting changes coming for companies. Here are some IoT trends and what people think will happen in the future:

Growth: It is thought that there will be tens of billions of Internet of Things devices in use in the next few decades. The number of IoT devices is expected to keep growing quickly. It will grow because more businesses will start to use it and because new use cases and applications will be made.

Computing at the edge: Edge computing is becoming more important for IoT because it lets data be handled and analyzed closer to where it comes from, instead of in a central data center. These changes can speed up responses, lower delays, and cut down on the quantity of data that needs to be sent over IoT networks.

AI and machine learning: AI and ML are becoming more and more important for IoT because they can look at the huge amounts of data that IoT devices produce and draw useful conclusions. This can help companies make better choices and run their businesses more efficiently.

Blockchain: Blockchain technology is being investigated as a way to make the Internet of Things safer and more private. IoT devices can connect to safe, open networks made with blockchain, which can reduce the risk of data security breaches.

Sustainability: As companies look for ways to have less of an effect on the world, sustainability is becoming more important for IoT. IoT can help many different types of businesses use energy more efficiently, cut down on waste, and become more environmentally friendly.

The future of IoT is exciting, with lots of new developments and innovations on the way. Device makers are also giving good deals, since it is getting cheaper to make IoT devices. As more IoT devices are added, companies will need to be ready to switch to new technologies and accept new uses and scenarios. People who can do this will be in a good situation to benefit from this important new technology.

Here are some examples of IoT applications that cater to the needs and preferences of individual citizens:

Home Security Enhancement: IoT-powered security systems that enable remote control and monitoring of alarm systems. These systems provide real-time alerts and notifications, contributing to the safety and security of the individual and their family members.

Activity Recognition for Elderly Care: IoT-based solutions that use sensors to recognize the activities and movements of older individuals in their homes. This technology can detect falls or unusual patterns of behavior, allowing for timely assistance and care.

Home Inventory Management Reminder: IoT applications that assist in managing household inventories. They can send reminders about expiring items, track usage, and even facilitate online shopping to replenish supplies when needed, making daily tasks more convenient.

These IoT applications are designed to enhance security, provide assistance, and simplify everyday tasks, ultimately improving the quality of life for individual citizens.

10.12 SUMMARY

In this chapter, we learned about the basics of the IoT, including its definition, architecture, and evolution over time. We also provide specific suggestions for the IoT's future uses. The IoT has the potential to improve a wide range of manufacturing, logistics, and food service appliances. Today's startups encounter increased difficulty in adapting to changing market conditions, regulatory frameworks, distribution channels, and consumer preferences. Therefore, many organizations rely on Industrial Internet of Things (IIoT).The term for any action taken by firms to model, monitor, and enhance their business operations by leveraging insights derived from a vast network of interconnected machines to enhance their economic profitability is commonly referred to as "IIoT."

REFERENCES

1. Nia A M and Jha N K 2017 A comprehensive study of the security of Internet-of-Things IEEE Trans. Emerging Top. Comput. 5 586–602
2. Daud M, Rasiah R and George M 2018 Denial of service: (DoS) impact on sensors 4th Int. Conf. on Info. Management (ICIM)
3. Yang Y, Wu L, Yin G, Li L and Zhao H 2017 A survey on security and privacy issues in internet-of-things IEEE Internet Things J. 4 1250–8
4. Jamal H, Huzaifa M and Sodunke M A 2019 Smart heat stress and toxic gases monitoring instrument with a developed graphical user interface using IoT Int. Conf. on Electrical, Commun., and Computer Engineering (ICECCE)
5. Kodali R K and Rajanarayanan S C 2019 IoT based indoor air quality monitoring system Int. Conf. on Wireless Commun. Signal Processing and Networking (WiSPNET)
6. Wu F, Wu T and Yuce M R 2019 Design and implementation of a wearable sensor network system for IoT-connected safety and health application IEEE 5th World Forum on the Internet of Things (WF-IoT)
7. Barman B K, Yadav S N and Kumar S 2018 IOT based smart energy meter for efficient energy utilization in smart grid 2nd Int. Conf. on Power, Energy and Environment: Towards Smart Technology (ICEPE)

8. Muthukumar S, Sherine Mary W and Jayanthi S 2018 IoT based air pollution monitoring and control system Int. Conf. on Inventive Res. in Comput. Appl. (ICIRCA)
9. Prabha B 2019 An IoT based efficient fire supervision monitoring and alerting system Third Int. Conf. on I-SMAC (IoT in Social, Mobile, Analytics and Cloud)
10. Agarwal K, Agarwal A and Misra G 2019 Review and performance analysis on wireless smart home and home automation using IoT Third Int. Conf. on I-SMAC.
11. Yang Y, Liu X and Deng R H 2018 Lightweight break-glass access control system for healthcare Internet-of-Things IEEE Trans. on Industrial Informatics 14 3610–7
12. Rajurkar C, Prabaharan S R S and Muthulakshmi S 2017 IoT based water management IEEE Int. Conf. on Nextgen Electronic Technologies: Silicon to Software (ICNETS2).

11 Leveraging Artificial Intelligence in Climate Change Interpretation

Overcoming Challenges in Risk Management Approach

Sugyanta Priyadarshini, Sarthak Dash and Sukanya Priyadarshini

11.1 INTRODUCTION

The anthropogenically driven climate change began with the inception of the Industrial era with its rampant use of machines (Oreskes, 2013; Vila-Traver et al., 2021) and heavy reliance on fossil fuels (Höök & Tang, 2013; Harnowo et al., 2021). Over the decades, climate change has accelerated (United Nations, 2020), resulting into the world viewing it as one of the biggest global challenges (Wagner, 2023; Uchiyama et al., 2020; Asadieh & Krakauer, 2017). According to the 2023 report of the United Nations' Intergovernmental Panel on Climate Change (IPCC), there is no doubt that human activities, particularly the emissions of greenhouse gases, are to blame for the increase in global surface temperature to 1.1°C higher during 2011–2020 than temperatures between 1850 and 1900. The IPCC concludes that, under all examined scenarios, there is a greater than 50% chance that the global temperature will rise by 1.5 degrees Celsius (2.7 degrees Fahrenheit) or more between 2021 and 2040. Under such a high-emissions trajectory, the globe may approach this threshold even sooner—between 2018 and 2037. This will in turn lead to disrupting ecological systems, diminishing habituality of earth, extinction of many species, economic losses, threat to peace and stability and spike in poverty (IPCC AR6 Synthesis report, 2023). Over the years, scientists, researchers and policy makers have developed many lofty goals, policies, adaptation and mitigation strategies to combat climate change. But all of these have failed to check its unprecedented rise. This epic fail gives the clarion call for the adoption of new technology to mitigate the transformational challenges induced by climate change. As the world has navigated its way into the Fourth Industrial Revolution (4IR) or digital revolution (World Economic Forum, 2016), innovative technologies like Artificial Intelligence (AI) hold immense potential to address the transformational challenges posed by climate change. The 4IR marks the upsurge of many technologies like AI (Jain et al., 2023), machine learning (Ravindiran et al., 2023), data analytics (Bhardwaj & Khaiter, 2023), blockchain (Cali et al., 2023), the Internet of things (IoT) (Alshahrani et al., 2023), cloud computing (Singh et al., 2021), quantum computing, advanced wireless communications and 3D printing. Among all these emerging technologies, AI has become ubiquitous in the contemporary world (Kurian et al., 2023; Cowls et al., 2023; Kovalishin et al., 2023). It has come a long way from conferences, debates and research with data scientists in the 1950s (Moor, 2006) to becoming a cutting-edge technology in today's society (Chatterjee et al., 2022). It can be defined as the science and engineering of creating intelligent devices, particularly intelligent computer programs (McCarthy, 2004; Kulkarni et al., 2023). It is a rapidly evolving field of computer science that deals with developing intelligent systems and deriving solutions to problems resembling human intelligence (Dignum, 2017). AI primarily aims to accelerate computer capabilities that are related to human understanding, reasoning, learning, problem-solving and predicting (Paschen et al., 2019; Osipov & Skryl, 2022). Along with the rapid developments in the field of AI, its subset, machine learning (ML) (Helm et al., 2020), has shown drastic advancements and is triggering breakthroughs in many research sectors, including climate change (Reichstein, 2019; Dijkstra et al., 2019). ML can be defined as an amalgamation of myriad technologies that depend on massive amounts of data to train an algorithm that can enable its continuous self-improvement (Meserole, 2018). With the data provided by the scientists and researchers, an educated guess can be made using ML, which can be further updated with more data feeding. Hence, the intersection of AI and ML holds tremendous potential in mitigating climate change.

However, mitigating climate change isn't a one-step process. It is a complex system of multiple variables like GHGs concentration, carbon dioxide level, precipitation level, wind pattern, fluctuating temperature and humidity level that makes it difficult to collect as well as analyze data and predict mitigation strategies. In such a scenario, AI seems particularly fit to address massive data challenges and face all queries in this field. AI holds paramount importance in making right decisions, framing apt government policies, programs and action plans (Kaack et al., 2021). By using large data, learning algorithms and sensing devices, AI has robust potential to evaluate weather change, forecast approaching catastrophes and reduce the magnitude of loss that might be created by such disasters. Additionally, using AI can also aid in establishing possibilities to escape undesired natural cataclysms and assist humans to make the right decision in such tough times. There are a plethora of papers that highlight the role of AI in almost every aspect of human life (Mohammad, 2020; Biswal, 2023; Cioffi et al., 2020; West & Allen, 2018) and the potential it holds in transforming every walk of life (Christakis, 2019). But there is a lacunae of research papers in understanding role of AI exclusively in climate change. Further, there are certain gaps and difficulties that prevent AI from reaching its full potential. Moreover, there is also a gap in literature regarding the numerous applications of AI in predicting and mitigating climate change. Given this scenario, this chapter attempts to draw attention towards (i) how AI can enhance our current knowledge and understanding of climate change, (ii) how AI can pose as an effective tool to speed up adaptation and mitigation strategies for climate change, (iii) how AI can provide sustainable and effective solutions to prevent undesired natural cataclysms and (iv) how AI plays a significant role in overcoming challenges in climatic risk management.

11.2 RESEARCH METHODOLOGY

This chapter has considered a systematic review methodology for the review conduction. In this review process, various steps have been incorporated, such as planning of review, search string and search criteria for artificial intelligence in climate change. After the completion of the search, the research articles are considered for review using inclusion and exclusion criteria. This section offers details on how the review is carried out.

11.2.1 REVIEW PLAN

In recent years, AI in climate change has come along substantially. Despite extensive study efforts, the prospective outcomes for each field have not been uncovered yet. This chapter has tried to conduct in-depth research and an overview of AI in the climate change domain. Further, this study has explored the implications of AI in various sub-categorical domains of climate change, including enhancing knowledge regarding climate change, adaptation and mitigation strategies of climate change, solutions in the prevention of undesired natural cataclysms and overcoming challenges in climatic risk management.

11.2.2 SEARCH STRING

In the extraction articles related to AI and climate change, several keywords are identified, such as AI, AI in climate change management, AI in the prediction of climatic variation and AI in the prevention and mitigation of climate change. However, the final data are extracted by applying specific keywords, which are "artificial intelligence", and "climate change". For data extraction, WOS and Scopus online databases are preferred, which are considered wide-ranging citation databases with the maximum number of peer-reviewed influential core, interdisciplinary and multidisciplinary journals. A total of 3783 articles (Scopus n=2091 and WOS n=1692) are extracted from the two online databases. However, some related articles may not have been considered because of the mismatch in title with the identified keywords. Figure 11.1 depicts the search string.

11.2.3 SELECTION CRITERIA

The review process opted for pre-specified selection criteria to include and exclude the papers. In the exclusion process, 222 papers from the Scopus database and 152 papers from the WOS database are excluded because of the unavailability of the DOI number of the papers. Further, 251 review papers and 780 duplicates are excluded.

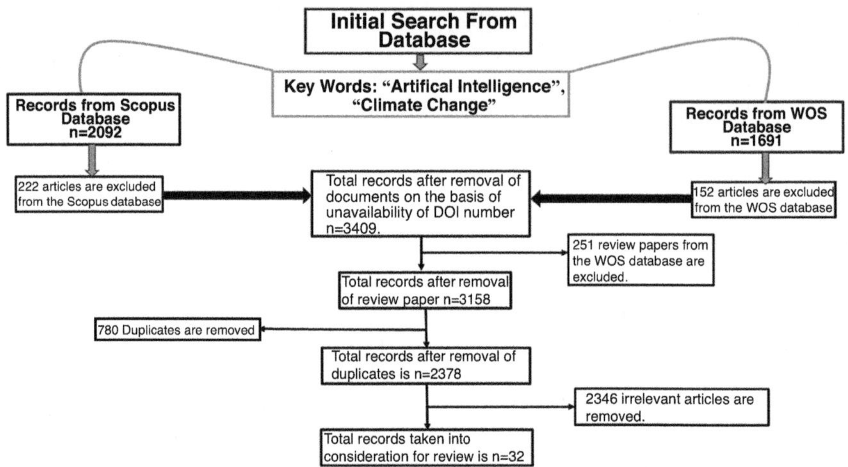

FIGURE 11.1　Search string.

Out of the remaining 2378 research articles, after thorough study 2346 articles are excluded as they are not relevant to the concerned categories mentioned in the earlier section, and a total of 32 articles are taken into consideration for the literature review presented in the figure.

11.3　DISCUSSION AND RESULTS

11.3.1　ROLE OF AI IN ENHANCING CURRENT KNOWLEDGE AND UNDERSTANDING OF CLIMATE CHANGE

Climate change is a complex problem. It is one of the greatest challenges faced by all of humanity (Rolnick et al., 2019). According to the 2023 IPCC Report, we are living in a climate emergency (IPCC AR6 Synthesis report, 2023). With its catastrophic effects on ecosystems, human civilizations and economic activity, climate change continues to rank among the most urgent global issues. Earlier discussions on climate change focused primarily on its physical aspects, like temperature rise, carbon dioxide levels, amount of rainfall, GHGs level and wind patterns. But no significant attention was being paid to it past these physical aspects. Behind these physical aspects lies the constantly changing data sets that determine the climate change. These data sets bring forth the role of AI in understanding climate change. Data is the primary source from which AI algorithms learn. They make decisions based on the data they are given, forecast relationships between variables, make predictions and assess their performance (Kaack et al., 2022). The performance of the algorithm improves with the quality of the data it learns from. Data is processed in a variety of ways for each individual application. It may be produced, assembled, pooled with other data, saved, labelled and deleted. A multitude of other technologies, inputs and procedures are used in conjunction with AI algorithms to ensure its success. These span from raw data to big data processing tools and platforms to computational infrastructure and a variety of more niche digital infrastructure.

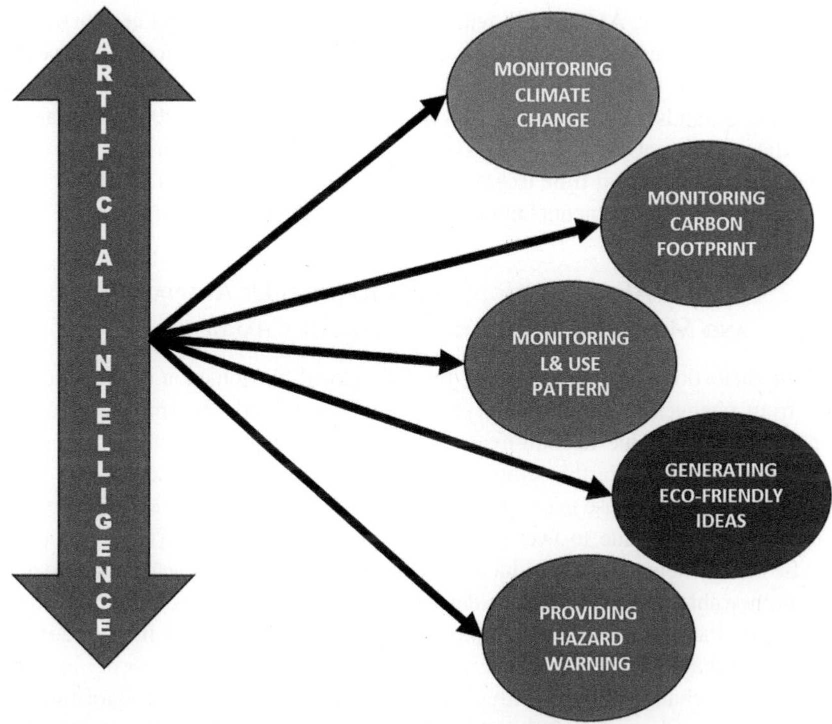

FIGURE 11.2   AI in enhancing current knowledge and understanding of climate change.

Although the full potential of AI in understanding and management of climate change is not fully explored yet, this developing area definitely has shown promising results (Sharifi & Amir Reza Khavarian-Garmsir, 2023). Climate change data sets are enormous and take significant time to collect, analyze and apply the data toward creating informed decisions, apt mitigation strategies and climate change policies (Ferreira et al., 2020). There has been a significant increase of carbon dioxide emission, rise in sea level, diminished crop productivity, constant biodiversity loss, increased frequency of natural disasters, changing weather patterns and amount of rainfall (Shivanna, 2022). As climate change is happening at a faster rate and the severity is rising, it is high time to implement innovative technology like AI in the climate change domain. With promising solutions posed by AI, it can be a panacea to mitigate climate change (Huntingford et al., 2019). Based on this, Figure 11.2 shows AI in Enhancing Current Knowledge and Understanding of Climate Change. The ability of artificial intelligence to interpret and gather data greatly reduces the discrepancy between forecasts made by digital models and actual circumstances, leading to more precise predictions of future outcomes (McGovern et al., 2017). Artificial intelligence is commonly utilized to search for all information and find new climate models, lowering forecast bias and enhancing accuracy (Jones, 2017). In recent years, AI has becoming more important in weather forecasting due to the vast amount of data provided by observation satellites and the complexity of climate models. Additionally, AI can help with the detection of regions that are particularly vulnerable to climate-related risks, the creation of adaption plans for businesses and communities, the forecasting of floods and wildfires and the detection of regions vulnerable to landslides (Sirmacek & Vinuesa, 2022; Bag et al., 2023). AI can also aid in the creation of early warning systems that can warn of imminent calamities, giving communities vital time to get prepared and evacuate faster (Rutenberg et al., 2021). Hence AI complements better understanding of climate change.

11.3.2 Role of AI as an Effective Tool to Speed Up Adaptation and Mitigation Strategies for Climate Change

Climatic distortions take place due to dynamic global phenomenon of climatic alteration from changes in temperature, precipitation and wind pattern. These rampant changes are caused due to man-made activities that result in global warming. The unprecedented rise in Earth's temperature caused by the greenhouse effect is one of the reasons for the increase in temperature and risk in climatic distortions. These issues remain unaddressed due to ever-growing urbanization and industrialization, which increase energy demand and production by burning of fossil fuels and over-utilization of non-renewable resources and can destruct the food chain and economic resources. Considering these possible devastations, AI plays a significant role in mitigating the impact of climate change distortions by predicting, analyzing, monitoring and mitigating climate change impacts with efficient use of data sets, learning algorithms and sensing devices (Chen et al., 2019). AI follows adaptation strategies such as data analysis and modelling, predictive modelling, risk assessment and vulnerability mapping, natural resource management, infrastructure mapping and design and follows mitigation strategies such as energy efficiency and optimization, renewable energy integration, smart grid management, emission monitoring and reporting, carbon pricing and market analysis, as represented in Figure 11.3. AI has already depicted its success in weather forecasting as well as environmental monitoring in several spatial locations.

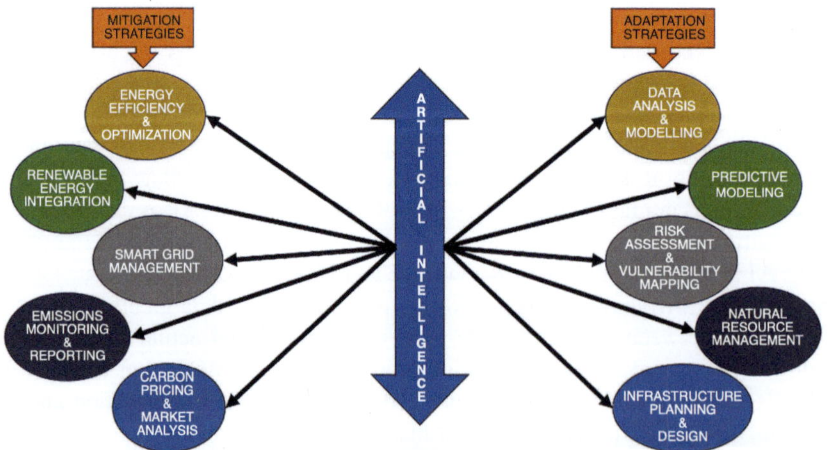

FIGURE 11.3 AI tool to speed up adaptation and mitigation strategies for climate change.

AI interprets daily weather events, climatic conditions and climatic data related to temperature and precipitation, and based on the technical mapping it predicts any possible distortions in the form of extreme weather. Along with predicting any distortion, it also alerts the enormous sources of carbon emissions so that the levels can be controlled beforehand (Sahil et al., 2023). This is intended to enable the policy makers to take steps to control the rising sea level, as well as extreme temperature and imposing carbon tax on the emitters. The Earth's hazards, like hurricanes, typhoons, species extinction, cyclone, drought and disruption of natural habitats, can also be predicted through AI operations. Climatic informatics has remained very helpful for the scientists and research experts in using AI paradigms in this approach. Unfortunately, the AI models used to date for prediction and interpretation are reliable for short-term forecasting but have diverged significantly from long-term prediction, assessment and mitigation. Thus, to mitigate climatic distortions, more advanced research is needed.

11.3.3   ROLE OF AI IN PROVIDING SUSTAINABLE AND EFFECTIVE SOLUTIONS TO PREVENT UNDESIRED NATURAL CATACLYSMS

AI is well-known for its ability to complete those tasks that were previously solely accomplished by humans. There would not be any question in considering climate change to be one of the most scientific complications that mankind has ever faced, as changes in climatic conditions are subjected to a complex system comprising an enormous number of variables, such as release of excessive carbon dioxide to the atmosphere (Fawzy et al., 2020), change in temperature and precipitation levels and wind patterns (Chen et al., 2019). As these climatic changes are not subjected to a fixed phenomenal pattern and keep constantly changing, it gets difficult for AI to predict any changes in the climatic condition beforehand. AI requires data sets of fixed patterns strategically collected in a specific time zone to reach valid conclusions to enact related policy changes. AI predicts upcoming climate changes based on prior data collected on particular trends and patterns. This requires designing AI models that would come up with the early warning systems, geo-spatial analysis, infrastructure resilience, disaster response and relief, ecosystem restoration, emergency communication (Yang et al., 2023), risk management and mitigation strategies and public awareness and education, as presented in Figure 11.4. These sustainable and effective solutions to prevent undesired natural cataclysms via AI can give the scientists better feedback about their running programmes, thereby improving the results of the existing models in predicting uncertainties in climate change (Sahil et al., 2023). AI has predominantly come up with tangible results. However, it has not come up with significant solutions in terms of water bodies, especially oceans. The ocean is one of the least predictable and understandable parts of the planet as it both transfers and absorbs heat. Furthermore, it is unknown how ocean bodies react to the changes occurring in the environment or in the climate. Additionally, to understand these changes, AI makes use of satellites revolving round the planet to make certain observations regarding climatic changes, such as monitoring forest fires and sources of carbon emission. With the growing and urgent need to predict climatic changes, a greater number of satellites are now in space to collect enormous information. For

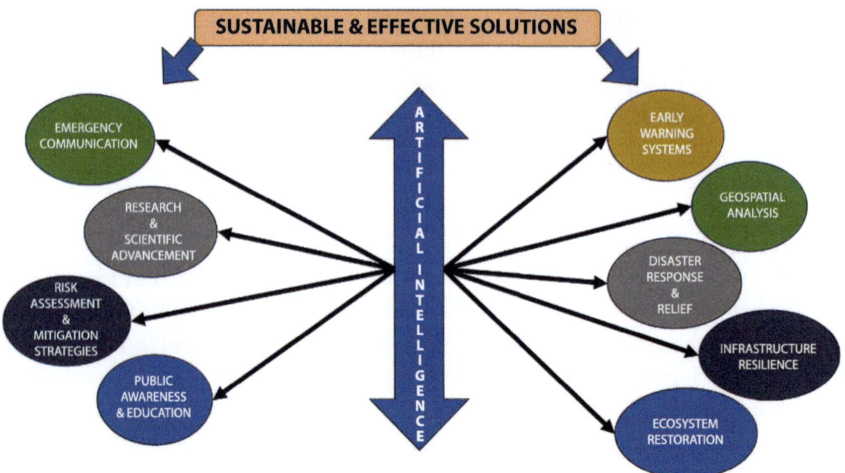

FIGURE 11.4 AI in sustainable and effective solutions to prevent undesired natural cataclysms.

instance, the Arctic Ocean is dramatically changing with the global warming touching unprecedented heights. To understand these changes, in summer and spring when the temperature is high, ships in the Arctic Ocean can collect the required data for AI to predict ocean data and make relevant changes in the policies for sustainable changes. But unfortunately, during winters as the temperature turns negative and the oceans turns to ice, the ships fail to enter the Arctic to conduct the operations, resulting in a data collection gap. In these situations, AI can be helpful in assimilating the required data to maintain continuity of data collection and predict the changes based on the given trends and patterns. This can be very helpful in predicting and analyzing the climatic distortions, if any, beforehand for sustainable and effective solutions.

11.3.4 Role of AI in Overcoming Challenges in Climatic Risk Management

AI plays a significant role in risk management in the context of global climate change by leveraging cutting-edge algorithms and machine learning techniques in order to detect, investigate and mitigate risks more effectively and efficiently with the help of patterns and correlations. These patterns and correlations can be helpful in prioritizing risks and allocating resources accordingly (Galaz et al., 2021).

11.3.4.1 AI and Climate Change Risk Identification

AI plays a crucial role in identifying possible risks associated with climate change by analyzing and interpreting enormous records of data from diverse sources. AI is widely used in augmented environmental monitoring (Hino et al., 2018) for effective climate mitigation action and robust farming practices, but it has yet to be developed prominently in terms of identifying potential risks. However, the converging trends in AI have come up with development and deployment of AI and associated technologies that can mitigate risks related to climate change from a sustainable perspective

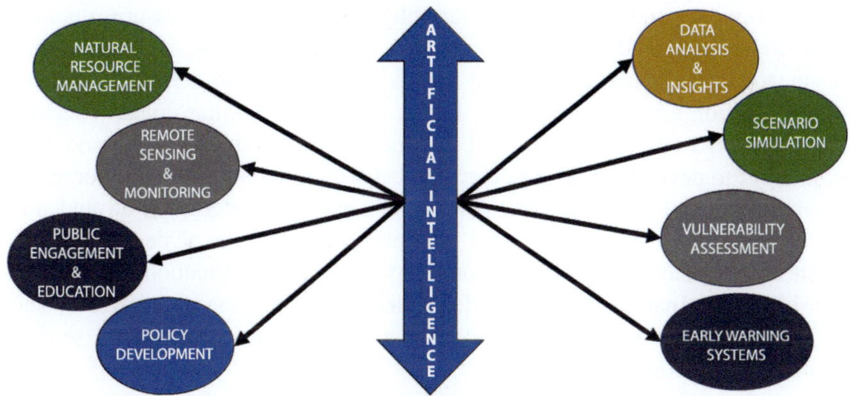

FIGURE 11.5   AI in overcoming challenges in climatic risk management.

(Dauvergne, 2020). AI can analyze troublemaking weather to identify any possible climatic disruption beforehand. For instance, AI algorithms can predict relationships between the unnatural behavior of any disaster such as a cyclone, hurricane or tsunami with climate factors like heat stored within the ocean, wind speed and air temperature to forecast any potential risk. Additionally, AI can interpret data from temperature and tectonics sensors to detect anomalies that could indicate a potential safety risk (Creutzig et al., 2019). In today's techno-world, a cyclone or storm landfall is usually predicted 48 hours beforehand. The AI-augmented model can be run before the disaster to determine probability so that warning can be issued to citizens who are at maximum risk and they can be moved to safer sites. The observational data collected from newly emerging AI, especially from automatic meteorological stations, radars and satellites has a very broad application in the context of meteorological big data (Zhou et al., 2022). The newly augmented AI tools, including deep learning, can understand complex structures to efficiently figure out disruptions in the meteorological field. Furthermore, AI's image recognition capability can come up with innovative ideas and directions for problematic climate monitoring and forecasting. Ultimately, the data collected by AI tools can be helpful not just for the organizations to trace potential risks to protect citizens from disaster, but also by providing valuable insights that can aid in undertaking hands-on measures to mitigate risks. Thus, it can be said that AI insights combined with human expertise can identify and mitigate climate-orientated disruptions by analyzing images and video data to identify potential risks.

11.3.4.2   AI and Climate Change Risk Assessment

Use of a convolutional neural network to estimate climatic disruption intensity based on satellite cloud pictures (Wimmers et al.,2019), use of an Artificial Neural Network in predicting tropical cyclones, use of a back-propagation (BP) algorithm to optimize the multilayer perceptron Multi-layer Perceptron (MLP) network for assessing climatic disruption (Baik & Paek, 2000) and use of a Multi-layer Perceptron to predict typhoon intensities (Chaudhuri et al., 2013) are some of the recent unique steps taken in AI approach to predict and mitigate climatic risks. The aforementioned works made use of variables such as central pressure, maximum sustained surface wind speed, pressure drop, total atmospheric ozone column and sea surface temperature as the input matrix of the model to determine the minimum prediction error of the model. Furthermore, in a similar context, recursive neural networks (RNN) can be used to demonstrate of temporal dynamic behavior. In order to develop an AI-based cyclonic disruption monitoring and forecasting system, automated and objective localization, intensity determination and intensity trend discrimination of cyclonic disruptions needs to be achieved, for which a vortex identification model based on deep image target detection, an intelligent intensity determination model based on image classification and retrieval and a fast enhancement discrimination model incorporating spatial-temporal sequence features are essential.

AI can be useful in analyzing risks by interpreting data and recognizing the patterns and correlations to predict the severity and probability of potential risks. For instance, the predictive models under AI augmented models can estimate the probability of the RNN based on prior data on previous breaches. AI can analyze risks by interpreting data and recognizing patterns and correlations by focussing on various risk scenarios. It can simulate the influence of a natural disaster on supply chain operations to help non-profit Sustainable Environment and Ecological Development Society (SEED) organizations prepare for and mitigate the risk. Further, AI can also help SEED organizations to assess the severity and likelihood of risks more precisely and proficiently by interpreting vast data and identify patterns and trends in prior existing data. AI plays a significant role in risk assessment, but it should not be the sole handler of the assessment. It should be accompanied by human expertise and judgement for effective risk management decisions.

11.4 PRACTICAL IMPLICATIONS

The aim of the research is not just to conduct a systematic literature review, but also to join the poorly connected strands in prior research domains that are widely related to the application of AI and implication of machine learning and robotics in sustainability climatic sciences, thereby guiding future researchers and policy makers to accelerate AI-based climate actions efficiently in terms of current policy debates to govern AI. The massive loss of life and property has remained a concern of mankind for more than 1,000 years. Voluminous in-depth research work has been conducted on major approaches for AI in climate disruptive domains, such as structure, dynamics and forecasting techniques that require large-scale data for detection, analysis and prediction. An innovative way of understanding how to address the bottlenecks in the field of climatic disruption via AI is the need of the hour. Technologies based on a purely data-driven approach will be difficult due to limited data availability for providing huge contributions to improving cyclonic disruption predictions. Even if the prior literature has made significant progress in genesis forecasts, path prediction, intensity prediction, weather prediction and improving numerical forecast models by integrating machine learning in AI, practical implication of these methodologies has stayed back and needs to be operated in the practical field, which can be regarded as both an opportunity and a challenge. Further, AI's potential in a climatic-oriented approach has been exploited only to a limited extent in the practical world due to unavailability or underutilization of the existing data. It is evident that the role of AI in leveraging a climatic-oriented approach is promising and challenging, which demands researchers and meteorologists have better knowledge and hand over climate-oriented dynamics to identifying key issues and solve them by developing suitable AI tools. The current research work can lay a foundation for further analysis and future studies related to AI tools in leveraging Artificial Intelligence in climate change interpretation.

11.5 CONCLUSION

The never-ending techno-driven human activity has substantially shaped the biosphere and climate system and thus the expectations of AI, and its tools, such as machine learning, deep learning, Internet of Things and robotics, have improved to detect, adapt and respond to climatic distortions beforehand in order to save lives and property. The vast diffusing urbanization of the 21st century has a significant impact on environmental changes, especially on the climate creating risk for Sustainable Development Goal 11, which centred on making human settlements more inclusive, safe, resilient and sustainable. As the data is unavailable or limited, usage of AI is restricted. It will be helpful if organizations such as Future Earth, the Global Carbon Project, International Council for Local Environmental Initiatives (ICLEI), Carbon Disclosure Project (CDP), Integrated Global Greenhouse Gas Information System (IG3IS), World Urban Database and Access Portal Tools (WUDAPT) and the C40 Network will come up with plenty of information to increase AI functionality. The aforementioned organizations can collaborate to create and curate a combined podium for financial data analysts, who can analyze and interpret all the data for effortless and effective AI functioning of. However, it is noteworthy that the existing models are learning methods, and the extreme climatic distortions cannot be predicted with 100% accuracy in the practical world. The need of the hour is to find the answers for how to build training datasets and build supervised AI tools to achieve predictive goals of climatic distortion prediction and mitigation.

REFERENCES

Alshahrani, A., Mahmoud, A., El-Sappagh, S., & Elbelkasy, M. S. A. (2023). An internet of things based air pollution detection device for mitigating climate changes. *Information Sciences Letters, 12*(4). *https://digitalcommons.aaru.edu.jo/isl/vol12/iss4/13*

Asadieh, B., & Krakauer, N. Y. (2017). Global change in streamflow extremes under climate change over the 21st century. *Hydrology and Earth System Sciences, 21*(11), 5863–5874. Scopus. *https://doi.org/10.5194/hess-21-5863-2017*

Bag, S., Sabbir Rahman, M., Rogers, H., Srivastava, G., & Harm Christiaan Pretorius, J. (2023). Climate change adaptation and disaster risk reduction in the garment industry supply chain network. *Transportation Research Part E: Logistics and Transportation Review, 171*, 103031. *https://doi.org/10.1016/j.tre.2023.103031*

Baik, J.-J., & Paek, J.-S. (2000). A neural network model for predicting typhoon intensity. *Journal of the Meteorological Society of Japan. Ser. II, 78*(6), 857–869.

Bhardwaj, E., & Khaiter, P. A. (2023). What data analytics can or cannot do for climate change studies: An inventory of interactive visual tools. *Ecological Informatics, 73,* 101918. *https://doi.org/10.1016/j.ecoinf.2022.101918*

Biswal, A. (2023). *Top 18 Artificial Intelligence (AI) Applications in 2023 | Simplilearn.* Simplilearn.Com. *www.simplilearn.com/tutorials/artificial-intelligence-tutorial/artificial-intelligence-applications*

Cali, U., Halden, U., Kuzlu, M., Pasetti, M., Gourisetti, S. N. G., Chandler, S., Rahimi, F., & Lima, C. (2023). Contribution of blockchain technology in energy to the climate change efforts. *2023 IEEE Power & Energy Society Innovative Smart Grid Technologies Conference (ISGT), 1–5. https://doi.org/10.1109/ISGT51731.2023.10066374*

Chatterjee, S., Chaudhuri, R., Kamble, S., Gupta, S., & Sivarajah, U. (2022). Adoption of artificial intelligence and cutting-edge technologies for production system sustainability: A moderator-mediation analysis. *Information Systems Frontiers. https://doi.org/10.1007/s10796-022-10317-x*

Chaudhuri, S., Dutta, D., Goswami, S., & Middey, A. (2013). Intensity forecast of tropical cyclones over North Indian Ocean using multilayer perceptron model: Skill and performance verification. *Natural Hazards, 65,* 97–113.

Chen, R., Wang, X., Zhang, W., Zhu, X., Li, A., & Yang, C. (2019). A hybrid CNN-LSTM model for typhoon formation forecasting. *GeoInformatica, 23,* 375–396.

Christakis, N. A. (2019). *How-AI-will-rewire-us.pdf. https://fully-human.org/wp-content/uploads/2019/09/How-AI-will-rewire-us.pdf*

Cioffi, Raffaele, Marta Travaglioni, Giuseppina Piscitelli, Antonella Petrillo, and Fabio De Felice. 2020. Artificial Intelligence and machine learning applications in smart production: Progress, trends, and directions. *Sustainability 12*(2), 492. https://doi.org/10.3390/su12020492

Cowls, J., Tsamados, A., Taddeo, M., & Floridi, L. (2023). The AI gambit: Leveraging artificial intelligence to combat climate change—opportunities, challenges, and recommendations. *AI & Society, 38*(1), 283–307. *https://doi.org/10.1007/s00146-021-01294-x*

Creutzig, F., Lohrey, S., Bai, X., Baklanov, A., Dawson, R., Dhakal, S., Lamb, W. F., McPhearson, T., Minx, J., Munoz, E., & Walsh, B. (2019). Upscaling urban data science for global climate solutions. *Global Sustainability, 2,* e2. *https://doi.org/10.1017/sus.2018.16*

Dauvergne, P. (2020). *AI in the Wild: Sustainability in the Age of Artificial Intelligence.* The MIT Press. *https://doi.org/10.7551/mitpress/12350.001.0001*

Dignum, V. (2017). *Responsible artificial intelligence: Designing AI for human values,* 1.

Dijkstra, H. A., Petersik, P., Hernández-García, E., & López, C. (2019). The application of machine learning techniques to improve El Niño prediction skill. *Frontiers in Physics, 7. www.frontiersin.org/articles/10.3389/fphy.2019.00153*

Fawzy, S., Osman, A., Doran, W., & Rooney, D. (2020). Strategies for mitigation of climate change: A review. *Environmental Chemistry Letters, 18. https://doi.org/10.1007/s10311-020-01059-w*

Ferreira, J. J. M., Fernandes, C. I., & Ferreira, F. A. F. (2020). Technology transfer, climate change mitigation, and environmental patent impact on sustainability and economic growth: A comparison of European countries. *Technological Forecasting and Social Change, 150,* 119770. *https://doi.org/10.1016/j.techfore.2019.119770*

Galaz, V., Centeno, M. A., Callahan, P. W., Causevic, A., Patterson, T., Brass, I., Baum, S., Farber, D., Fischer, J., Garcia, D., McPhearson, T., Jimenez, D., King, B., Larcey, P., & Levy, K. (2021). Artificial intelligence, systemic risks, and sustainability. *Technology in Society, 67,* 101741. *https://doi.org/10.1016/j.techsoc.2021.101741*

Harnowo, D., Indriani, F. C., Susanto, G. W. A., Prayogo, Y., & Mejaya, I. M. J. (2021). Biodiversity conservation through sustainable agriculture, its relevanve to climate change: A review on Indonesia situation. *Scopus, 911*(1). *https://doi.org/10.1088/1755-1315/911/1/012066*

Helm, J. M., Swiergosz, A. M., Haeberle, H. S., Karnuta, J. M., Schaffer, J. L., Krebs, V. E., Spitzer, A. I., & Ramkumar, P. N. (2020). Machine learning and artificial intelligence: Definitions, applications, and future directions. *Current Reviews in Musculoskeletal Medicine*, *13*(1), 69–76. *https://doi.org/10.1007/s12178-020-09600-8*

Hino, M., Benami, E., & Brooks, N. (2018). Machine learning for environmental monitoring. *Nature Sustainability*, *1*(10), Article 10. *https://doi.org/10.1038/s41893-018-0142-9*

Höök, M., & Tang, X. (2013). Depletion of fossil fuels and anthropogenic climate change—A review. *Energy Policy*, *52*, 797–809. *https://doi.org/10.1016/j.enpol.2012.10.046*

Huntingford, C., Jeffers, E. S., Bonsall, M. B., Christensen, H. M., Lees, T., & Yang, H. (2019). Machine learning and artificial intelligence to aid climate change research and preparedness. *Environmental Research Letters*, *14*(12), 124007. *https://doi. org/10.1088/1748-9326/ab4e55*

IPCC AR6 Synthesis report. (2023). *AR6 Synthesis Report: Climate Change 2023 — IPCC. www.ipcc.ch/report/sixth-assessment-report-cycle/*

Jain, H., Dhupper, R., Shrivastava, A., Kumar, D., & Kumari, M. (2023). AI-enabled strategies for climate change adaptation: Protecting communities, infrastructure, and businesses from the impacts of climate change. *Computational Urban Science*, *3*(1), 25. *https://doi. org/10.1007/s43762-023-00100-2*

Jones, N. (2017). How machine learning could help to improve climate forecasts. *Nature*, *548*(7668), Article 7668. *https://doi.org/10.1038/548379a*

Kaack, L. H., Donti, P. L., Strubell, E., Kamiya, G., Creutzig, F., & Rolnick, D. (2022). Aligning artificial intelligence with climate change mitigation. *Nature Climate Change*, *12*(6), Article 6. *https://doi.org/10.1038/s41558-022-01377-7*

Karl, T. R., & Trenberth, K. E. (2003). Modern global climate change. *Science*, *302*(5651), 1719–1723. *https://doi.org/10.1126/science.1090228*

Kovalishin, P., Nikitakos, N., Svilicic, B., Zhang, J., Nikishin, A., Dalaklis, D., Kharitonov, M., & Stefanakou, A.-A. (2023). Using Artificial Intelligence (AI) methods for effectively responding to climate change at marine ports. *Journal of International Maritime Safety, Environmental Affairs, and Shipping*, *7*(1), 2186589. *https://doi.org/10.1080/25725084 .2023.2186589*

Kulkarni, G. R., Tamta, M. K., Kumar, A., Nomani, M. Z. M., Singh, C., & Pallathadka, H. (2023). The role of artificial intelligence (AI) in creating smart energy infrastructure for the next generation and protection climate change. In S. Yadav, A. Haleem, P. K. Arora, & H. Kumar (Eds.), *Proceedings of Second International Conference in Mechanical and Energy Technology* (pp. 457–464). Springer Nature. *https://doi. org/10.1007/978-981-19-0108-9_47*

Kurian, N., Cherian, J. M., Sudharson, N. A., Varghese, K. G., & Wadhwa, S. (2023). AI is now everywhere. *British Dental Journal*, *234*(2), Article 2. *https://doi.org/10.1038/ s41415-023-5461-1*

Lynn H Kaack, Priya L Donti, Emma Strubell, George Kamiya, Felix Creutzig, et al.. Aligning artificial intelligence with climate change mitigation. 2021. ffhal-03368037.

McCarthy, J. (2004). *What is Artificial Intelligence?* http://jmc.stanford.edu/articles/whatisai/ whatisai.pdf

McGovern, A., Elmore, K. L., Gagne, D. J., Haupt, S. E., Karstens, C. D., Lagerquist, R., Smith, T., & Williams, J. K. (2017). Using Artificial Intelligence to Improve Real-Time Decision-Making for High-Impact Weather. *Bulletin of the American Meteorological Society*, *98*(10), 2073–2090. *https://doi.org/10.1175/BAMS-D-16-0123.1*

Meserole, C. (2018). *What is machine learning?* Brookings. *www.brookings.edu/articles/ what-is-machine-learning/*

Mohammad, S. M. (2020). AI automation and application in diverse sectors. *SSRN Electronic Journal*, *68*, 75–81. *https://doi.org/10.14445/22312803/IJCTT-V68I1P116*

Moor, J. (2006). The Dartmouth college artificial intelligence conference: The next fifty years. *AI Magazine*, *27*(4), Article 4. *https://doi.org/10.1609/aimag.v27i4.1911*

Oreskes, N. (2013). On the "reality" and reality of anthropogenic climate change. *Climatic Change, 119*(3), 559–560. *https://doi.org/10.1007/s10584-013-0779-3*

Osipov, V. S., & Skryl, T. V. (2022). AI's contribution to combating climate change and achieving environmental justice in the global economy. *Frontiers in Environmental Science, 10. www.frontiersin.org/articles/10.3389/fenvs.2022.952695*

Paschen, U., Pitt, C., & Kietzmann, J. (2019). Artificial intelligence: Building blocks and an innovation typology. *Business Horizons, 63. https://doi.org/10.1016/j.bushor.2019.10.004*

Ravindiran, G., Rajamanickam, S., Kanagarathinam, K., Hayder, G., Janardhan, G., Arunkumar, P., Arunachalam, S., AlObaid, A. A., Warad, I., & Muniasamy, S. K. (2023). Impact of air pollutants on climate change and prediction of air quality index using machine learning models. *Environmental Research, 239,* 117354. *https://doi.org/10.1016/j. envres.2023.117354*

Reichstein, M. (2019). *Deep learning and process understanding for data-driven Earth system science.* Nature. *www.nature.com/articles/s41586-019-0912-1*

Rolnick, D., Donti, P. L., Kaack, L. H., Kochanski, K., Lacoste, A., Sankaran, K., Ross, A. S., Milojevic-Dupont, N., Jaques, N., Waldman-Brown, A., Luccioni, A., Maharaj, T., Sherwin, E. D., Mukkavilli, S. K., Kording, K. P., Gomes, C., Ng, A. Y., Hassabis, D., Platt, J. C., . . . Bengio, Y. (2019). *Tackling Climate Change with Machine Learning* (arXiv:1906.05433). arXiv. *https://doi.org/10.48550/arXiv.1906.05433*

Rutenberg, I., Gwagwa, A., & Omino, M. (2021). Use and impact of artificial intelligence on climate change adaptation in Africa. In N. Oguge, D. Ayal, L. Adeleke, & I. da Silva (Eds.), *African Handbook of Climate Change Adaptation* (pp. 1107–1126). Springer International Publishing. *https://doi.org/10.1007/978-3-030-45106-6_80*

Sahil, K., Mehta, P., Bhardwaj, S., & Dhaliwal, L. (2023). *Development of Mitigation Strategies for the Climate Change Using Artificial Intelligence to Attain Sustainability* (pp. 421–448). Springer International Publishing. *https://doi.org/10.1016/ B978-0-323-99714-0.00021-2*

Sharifi, A., & Khavarian-Garmsir, A. R. (2023). *Urban Climate Adaptation and Mitigation—1st Edition.* Elsevier. *https://shop.elsevier.com/books/urban-climate-adaptation- and-mitigation/sharifi/978-0-323-85552-5*

Shivanna, K. R. (2022). Climate change and its impact on biodiversity and human welfare. *Proceedings of the Indian National Science Academy, 88*(2), 160–171. *https://doi. org/10.1007/s43538-022-00073-6*

Singh, M., Tuli, S., Butcher, R. J., Kaur, R., & Gill, S. S. (2021). Dynamic shift from cloud computing to industry 4.0: Eco-friendly choice or climate change threat. In P. Krause & F. Xhafa (Eds.), *IoT-based Intelligent Modelling for Environmental and Ecological Engineering: IoT Next Generation EcoAgro Systems* (pp. 275–293). Springer International Publishing. *https://doi.org/10.1007/978-3-030-71172-6_12*

Sirmacek, B., & Vinuesa, R. (2022). Remote sensing and AI for building climate adaptation applications. *Results in Engineering, 15,* 100524. *https://doi.org/10.1016/j. rineng.2022.100524*

Uchiyama, C., Stevenson, L. A., & Tandoko, E. (2020). Climate change research in Asia: A knowledge synthesis of Asia-Pacific Network for Global Change Research (2013–2018). *Environmental Research, 188. https://doi.org/10.1016/j.envres.2020.109635*

United Nations. (2020). *The Climate Crisis—A Race We Can Win.* United Nations. *www. un.org/en/un75/climate-crisis-race-we-can-win*

Vila-Traver, J., Aguilera, E., Infante-Amate, J., & González de Molina, M. (2021). Climate change and industrialization as the main drivers of Spanish agriculture water stress. *Science of the Total Environment, 760. https://doi.org/10.1016/j.scitotenv.2020.143399*

Wagner, P. (2023). The triple problem displacement: Climate change and the politics of the Great Acceleration. *European Journal of Social Theory, 26*(1), 24–47. *https://doi. org/10.1177/13684310221136083*

West, D. M., & Allen, J. (2018). *How artificial intelligence is transforming the world.* Brookings. *www.brookings.edu/articles/how-artificial-intelligence-is-transforming-the-world/*

Wimmers, A., Velden, C., & Cossuth, J. H. (2019). Using deep learning to estimate tropical cyclone intensity from satellite passive microwave imagery. *Monthly Weather Review, 147*(6), 2261–2282. *https://doi.org/10.1175/MWR-D-18-0391.1*

World Economic Forum. (2016, January 14). *The fourth industrial revolution: What it means and how to respond.* World Economic Forum. *www.weforum.org/agenda/2016/01/the-fourth-industrial-revolution-what-it-means-and-how-to-respond/*

Yalkı, İ. (2023). Industrialization impact on climate change: An examination of NICs. In *Handbook of Research on Sustainable Consumption and Production for Greener Economies* (pp. 1–26). IGI Global. *https://doi.org/10.4018/978-1-6684-8969-7.ch001*

Yang, M., Chen, L., Wang, J., Msigwa, G., Osman, A. I., Fawzy, S., Rooney, D. W., & Yap, P.-S. (2023). Circular economy strategies for combating climate change and other environmental issues. *Environmental Chemistry Letters, 21*(1), 55–80. *https://doi.org/10.1007/s10311-022-01499-6*

Zhou, G., Fang, X., Qian, Q., Lv, X., Cao, J., & Jiang, Y. (2022). Application of artificial intelligence technology in typhoon monitoring and forecasting. *Frontiers in Earth Science, 10.* *www.frontiersin.org/articles/10.3389/feart.2022.974497*

Index